COPE'S plastics BOOK

By

DWIGHT COPE

PRESIDENT, COPE PLASTICS

Edited by

FLOYD DICKEY

HEAD, INDUSTRIAL EDUCATION DEPARTMENT
JOHN ADAMS H. S., SOUTH BEND, IND.

South Holland, Illinois

THE GOODHEART-WILLCOX CO., INC.

Publishers

Copyright 1973

By

THE GOODHEART-WILLCOX CO., INC.

Previous Editions Copyright
1957, 1960

No part of this book may be reproduced in any form
without violating the copyright law. Printed in U.S.A.
Library of Congress Catalog Card Number 60-8667.
International Standard Book Number 0-87006-150-X.

23456789-73-987

INTRODUCTION

Cope's PLASTICS BOOK tells and shows how to make more than 90 useful and colorful Craft Projects from the modern wonder materials known as Plastics.

The projects range from Elementary to Advanced, and are suitable for both School and Home Workshop use.

Cope's Plastics Book describes various types of plastics and tells how they are made. It describes the best methods of working with Plastics--Storing, Sawing, Machining, Polishing, Heat Forming, Cementing, Bonding and Cleaning. There are special sections on Internal Carving and Dyeing, and Commercial Fabrication Methods.

It is hoped that the book will serve still another purpose--help those who use it, develop an appreciation of the importance of plastics in our daily living, as well as the great possibilities the plastics industry offers for the future.

The authors wish to extend here, their deep appreciation to the many organizations and individuals who have helped in the development of the book. Special credit is due:

American Cyanamid Co., New York, N.Y.
Castolite Co., Woodstock, Ill.
Chicago Wheel & Mfg. Co., Chicago, Ill.
Foredom Electric Co., New York, N.Y.
Grieve-Hendry Co., Chicago, Ill.
Hazelwood Brothers, Salina, Kans.
Hotpack Electric Co., Philadelphia, Pa.
Moslo Machinery Co., Cleveland, Ohio
O'Neill-Irwin Mfg. Co., Lake City, Minn.
Pasadena Hydraulics, Inc., El Monte, Calif.
Plastic Parts & Sales Co., St. Louis, Mo.
Polymer Corp., Reading, Pa.
Kelly Powell Photos, Chicago, Ill.
Rohm and Haas Co., Philadelphia, Pa.
Society of the Plastics Industry, Inc.
F. J. Stokes Co., Philadelphia, Pa.
Ray Twillman Pattern Co., St. Louis, Mo.
W. Dwight Meyer, for his invaluable assistance in the chemical terminology and preparation of this book.
Floyd Dickey, who spent many hours editing the book, and to his students who made the fine drawings.
Mozelle Cope, for the many hours spent in assisting with the preparation of the book.
The many Instructors in Industrial Arts, Craftsmen, and Friends who contributed projects.

CONTENTS

FACTS ABOUT PLASTICS 9
Definition of Plastics - First Synthetic Plastic - Organic and Inorganic Substances - Characteristics of Organic Substances - Polymerization - Forming New Substances - Catalyst - Classifying Plastics - Thermosetting Plastics - Thermoplastic Materials - Acrylic Plastics.

TYPES OF PLASTICS IN COMMON USE 14
The Phenolics - The Ureas - The Melamines - The Polyesters - Epoxy Resins - The Silicones - The Acrylics - Lighting Characteristics of Acrylics - Polystrene - Polyethylene - Polyamides - Vinyl Plastics - Polyvinylidene Chloride Plastics - Synthetic Rubbers - Polyfluoro Hydrocarbons - Other Plastic-Like Materials - The Cellulosics - Natural Resins - Shellac - Casein - Lignin.

WORKING WITH ACRYLICS 31
Forms in Which Plastics Come - Storing Plastic Stock - Transferring Designs to Mateial - Sawing Plastics - Machining Acrylics - Drilling in Plastics - Threading and Tapping - Sanding Plastics - Buffing and Polishing Plastics - Solvent Polishing - Heat Forming - The Strip Heater - Cementing and Bonding - Using Jigs - Heat Welding - Friction Welding - Cleaning - Antistatic Treatment - Waxing.

PROJECTS FROM PLASTICS 55
Tooth Brush Holder - Plastic Pie, Cake Servers - Nut Tray - Tie Slides - Styrofoam Ornaments - Plastic Finger Rings - Arm Bracelets - Pencil Holder - Styrofoam Cutouts - Styrofoam Fireplace Scene - Lampshades from Polyplastex - Napkin Holder - Letter and Note Holder - Laminated Letter Openers - Pen and Letter Holder - Plexiglas Bud Vase - Powder Box with Octagon Base - Perfume Vial Holders - Fish-Shaped Coin Bank - Tie Rack - Letter Opener - Plastic Cup with Cover - Candle Holder - Candy Dish from Patterned Surface Plastic - Coin Holders - Bud Vase - Fluted Candle Holders - Internally Carved Pin and Earring Set - Cube Lighters - Pair of Candle Holders - Magazine Rack - Picture Frame - Tenite Tool Handles - Harp-Shaped Earring Rack - Napkin Holder - Musical Clef Decorative Shelf - Pipe Holder - Wheelbarrow Candy, Nut Tray - Pin-Up Lamp - Clef Bud Vase - Desk Lighters - Carved-Lid Jewel Box - Wall Towel Rack - Compressed Cube Carving - Laminated Block Perfume Bottle Holders - Laminated Pen Sets - Raised Letter Name Plate - Stand for Electric Clock - Vanity Mirror - Knick-Knack Shelf - Hot Dish Pad - Bird Feeder - Stand for Rubber Stamps - Tie Rack of Clear Plastic - Heart-Shaped Perfume Atomizer Stand - White Swan Candy Dish - Planter Lamp - Swept-Corner Box - Three Tier Tray - Salt and Pepper Shakers - Cattail Centerpiece - Cribbage Board - Blue and White Candy Box - Black and White Jewel Box - Table Lamp - Cradled Jewel Box - Checkerboard - Sliding Lid Jewel Box - Lapel Pins - Three-cone Table Lamp - Night or TV Lamp - TV Lamp of Modern Design - Modern Free-Form Lamp - Plastic Rod Lamp - Torchere Lamp - Lazy Susan - Laminated Gavel - Cigarette Box and Dispenser - Musical Piano Cigarette Dispenser, or Jewel Box.

Contents

INTERNAL CARVING AND COLORING OF ACRYLICS......230
Basic Requirements - First Step - Controlling Cutting Tool - Special Drills Recommended - Plastic Should Be Polished - Carving Roses - Carving Other Designs - Using Dyes - Blending Granules - Carving Thin Stock - Jewelry Findings - Laminating Dyes - Painting Acrylics - Sculpturing Plastics.

POLYESTER RESINS AND REINFORCED LAMINATES......242
Properties of Laminates - Fiberglas - Polyester Resins - Wet Lay-Up Method of Plastic Lamination - Applying Resin with Brush - Applying Plastics to Metals - Fiberglas-Resin Serving Bowl - Overnight Case - Fiberglas-Polyester Resin Chair.

OTHER APPLICATIONS OF PLASTICS (COMMERCIAL)......255
Types of Plastic Moldings - Molding Procedure - Compression Molding - Injection Molding - Extrusion Molding - Built-Up Laminates - Color Laminates - Plastic Casting - Cast Phenolics - Other Resin Casting - Castolite Casting - Plastic Foams - Sheets and Films - Filaments - Metallizing Plastics.

GLOSSARY OF TERMS......264

INDEX......269

SOME FACTS ABOUT PLASTICS

The dictionary tells us that plastic means a material that is capable of being molded or shaped to any desired form.

Clay was one of the earlier forms of "plastic" material. It was shaped to the desired form and baked or fired to set it. Prehistoric man fashioned pottery in this manner. Pottery samples are among the oldest relics we have found of early man.

Another ancient plastic material was asphaltum or asphalt. This was found in natural deposits near old oil seeps, and is the "pitch" mentioned in the Bible. It was used thousands of years ago to caulk cracks between the boards in boats, and various other purposes.

Under this same definition of plastic we might think of metals that have been softened by heating as being plastic. Steel, heated in an old-fashioned forge becomes sufficiently plastic to be shaped by a few blows with a hammer. If heated to the flowing point, it may be poured into molds and cast to the desired shape. In our modern use of the word plastic a new meaning has appeared. This meaning is fast growing in popular usage and is applied to the growing list of synthetic substances that have been introduced to the world of manufacturing and engineering through the magic of modern chemistry.

Synthetic literally means built up by putting together the various parts. To the chemist the word synthetic simply means that the product was built up by using little chemical "building blocks" which we call atoms and molecules, and putting them together in the proper manner to produce the substance desired. In the laboratory it is a simple matter to combine the two substances hydrogen and oxygen to produce water. The water thus formed is like water we drink, only purer. The modern chemist can take various materials and combine them in such a way as to produce quinine that is the same as quinine extracted from cinchona bark. The chemist's product is pure quinine--even though it has been "put together" in the laboratory. It is therefore "synthetic." In our discussion of plastics we are going to use "synthetic" in a little narrower meaning. The product we refer to as synthetic is one that does not occur naturally, but has been developed by chemists, by putting together materials that are not combined in any of the "natural" processes.

FIRST SYNTHETIC PLASTIC

The first member of our modern family of synthetic materials called Plastics appeared on the scene in the early 1900's. A chemist named Leo H. Baekeland found that when he mixed ordinary carbolic acid (chemical name, phenol) with formaldehyde, he always got some gummy, unmanageable stuff that nothing would dissolve or wash out of his test tubes and other laboratory apparatus. He reasoned that anything as nearly indestructible as this gummy stuff should have important uses if he could only make the material behave.

After much experimenting with the reaction between phenol and formaldehyde, Dr. Baekeland finally found a method of making a material that could be molded into any desired shape, and "set" by further heating and compressing. This first synthetic "resin" was named Bakelite in his honor.

This discovery by Baekeland started a veritable army of chemists on the search for other materials that could be combined in the same way as phenol and formaldehyde were in Bakelite. In the beginning the progress was very slow, with plenty of setbacks. When some of the efforts succeeded, there were more and more people trying to find other combinations of materials or other chemical reactions. Today there are so many different kinds of "plastic" materials and so many variations of each basic type of material that it would be foolish to try to even mention all of them here. In the next chapter we shall list some of the more common basic types of plastics, and explain something about them.

ORGANIC AND INORGANIC SUBSTANCES

There are many things about these materials we can discuss in general terms. Practically all modern plastics are synthetic--in the sense that they are man-made, and do not occur in nature. All of them belong to the class we call organic substances. This term also needs to be defined, because it will give us certain clues to the over-all behavior of plastics.

When the science of chemistry was in its infancy, the men working at it discovered what appeared to be two entirely different kinds of matter. One kind could be dug up out of the earth, or obtained from other inanimate sources. The other kind, was found only in plant or animal matter. They reasoned that this latter kind of materials could be made only with the help of some living organism--either plant or animal--and so they called these materials organic matter. Many years later chemists discovered ways of making many so-called organic substances in the laboratory without the help of any living organism. By this time the division of chemistry into organic and inorganic had been so widely adopted that it would have been foolish to try to change it.

Early work with the so-called organics showed that they contained the element Carbon (represented by the chemical symbol "C"). In order to make the division of chemicals into two types appear logical, it was ruled that the study of carbon compounds be called organic chemistry.

Since the first laboratory synthesis of a true organic compound (the substance known as urea, made in the laboratory a little over 100 years ago) chemists have built up hundreds of thousands of carbon-containing substances. There are so many of these chemicals known today that a mere list of them would fill a large book.

CHARACTERISTICS OF ORGANIC SUBSTANCES

Let us consider some of the general characteristics of organic, or carbon-containing substances. We know that a molecule of any substance is the smallest possible piece of that substance that still retains all the properties of the parent substance. Each molecule is made up of one or more atoms. As an example of molecules and atoms, let's consider ordinary water. It is easy to find out that water is made up of two different kinds of atoms--oxygen atoms and hydrogen atoms. It is also easy to determine that there are two hydrogens (H) for every oxygen (O). We can imagine a molecule of water as being made up like this H-O-H--that is, with two hydrogens "bonded" or hanging onto the oxygen. That would be the smallest possible amount of water we could have, because if this tiny bit lost one of its hydrogens, it would no longer be water.

Most of our inorganic molecules are quite simple in their make-up or structure. These molecules are made up of only two, three, or four different kinds of atoms, and with only a relatively few atoms to the molecule. Organic substances are a different matter. They can be unbelievably large and complex. This is because the carbon atom has the habit of combining with other carbons, as well as with atoms of other elements. In this fashion, carbons can combine into long chains, or into rings, with atoms of other elements hanging onto the sides.

If we could magnify a molecule of a carbon-chain substance so we could see it, it might look like a long caterpillar. The backbone of the caterpillar would be represented by the carbon atoms, hooked to each other forming a long chain, while the legs of the caterpillar would correspond to the other atoms hanging onto the sides of the carbons. If the caterpillar could swing around and take its tail in its mouth, it would form a ring-like structure. In the case of carbon compounds we find many ring-like molecules, and many instances of a special type of ring containing six carbon atoms. This particular ring is known as the benzene ring, and compounds containing this type of ring form a separate series of carbon compounds called the aromatics, because many of them have a decided and typical aroma or smell. The straight chain-like molecules belong to the aliphatic series.

In addition to the ring-type and the chain-type carbon compounds, it is possible for short, or even long, chains to attach to rings, or to other chains so as to make branched-chain compounds. Molecules of this latter type may look like the branched twig of a tree. Keep in mind this picture of chain-like and ring-like molecules with a carbon backbone, because we shall refer to them later.

An interesting thing about carbon compounds is that the carbon atoms cling to each other, and also to other kinds of atoms, with varying degrees of force. The strength of this clinging force, or bond as the chemists call it, varies greatly. The variation may be caused by the presence of other atoms. It is also true that in most cases this bond becomes weaker as the temperature gets higher. For this reason most organic substances are unstable (that is, break up easily) when heated. Most carbon compounds will char or burn if they are put in an ordinary kitchen oven at the temperature it would take to bake an apple pie. Since most of our plastics are organic in nature, they are also sensitive to heat. This makes them unsuitable when used at high temperatures. This must be taken into consideration when we are thinking of ways to use them.

POLYMERIZATION

Another basic idea we must understand before we start to talk about plastics is contained in the chemist's polymer and polymerize. The "poly" means "many"--the "mer" is from the Greek word "meros," meaning "part." In other words a polymer is a substance whose molecules are composed of many parts. Polymerization is the act of combining two or more molecules into a single large molecule, thus creating a different substance. This new substance will have different properties from the original substance. This is because every time you change the structure of a molecule, you form another and different substance.

FORMING NEW SUBSTANCES

From this explanation we can see that polymerization is a method of forming new substances--provided we can get the molecules to combine with each other, and can control the combining operation. Some molecules cannot be induced to combine, no matter what we do. Others can be made to combine by adding a bit of a special substance called a catalyst. In many cases, once the molecules get the idea of combining, they combine very rapidly and we find it difficult to get them to stop at the proper place.

If we have the molecules of a polymer-making substance that have not yet done any combining, it is called a monomer. If the combining action is stopped as soon as two molecules have combined into one, we call the new product a dimer--if three, it is called a trimer. In most cases, after the combining has been started, a number of the smaller molecules join up to form a large molecule and in this case the new molecule is called a polymer. Some of the molecules of polymers are very huge as compared with the simpler molecules of the monomer with which the plastics maker starts his work. The polymerized molecule may contain several thousand of the molecules of the original monomer.

Not all molecules can be made to polymerize. Usually the molecules that have this tendency to unite with others have some point of weakness. One type of weak spot is what the chemists call a double or "unsaturated" bond between carbon atoms. We can explain this by taking one of the simplest compounds of carbon called ethane. It looks something like this:

$$\begin{array}{c} H\ H \\ H-C-C-H \\ H\ H \end{array}$$

In this molecule each of the carbons has one of its four normal bonds or holding arms attached to a hydrogen or to another carbon. If we were to knock two hydrogens off this molecule it would have a form like this:

$$\begin{array}{c} H-C-C-H \\ H\ H \end{array}$$

This substance is called ethylene. Here the carbons are held together by two bonds. Strangely, this is not as stable a condition as when there is only one bond between the carbons. One of the bond lines, representing an arm from each of the two carbons, wants to break up--and will do so at the first reasonable opportunity. Under the proper conditions this unsaturated bond--it's called that because it does not hold all the hydrogens or other atoms it could--breaks loose and grabs onto anything in the neighborhood that it can find. Suppose there were two of the nervous ethylene molecules close together and some outside force caused this weak bond to open up. Then the two could unite at that point to form a new molecule. In this case, we might call it di-ethylene, or the dimer of ethylene. Under the usual conditions of treatment of ethylene, a large number of the ethylenes would combine to form a huge molecule of polyethylene. This is one of our favorite plastic materials, as we shall see later.

Other conditions in a molecule besides an unsaturated bond may offer a place where one of the bonding arms of the molecule will shake loose from what it is holding onto, and grab a nearby molecule. If this process continues from molecule to molecule we again have a set of conditions that will allow the building up of a large polymerized molecule. An example of this is the phenol molecule (we called it carbolic acid earlier). This is a ring-like substance with an OH group on it that is not so tightly held as the others and wants to go wandering. This leaves the free arm ready to grab onto anything that comes along. If there are a lot of phenol molecules and also a number of formaldehyde molecules together, they will join to form a large phenol-formaldehyde group. This is the basic resinous molecule first brought into the modern plastics group by Dr. Baekeland.

There are a number of other groupings of atoms at certain points in a molecule that are jittery under the proper conditions, and that will break up or rearrange so that free bonds are left to grab onto other molecules. Any arrangement of a group of atoms in a molecule that has this jittery tendency is called a "functional" group. It is the point of attack-- the point where other molecules can join onto it to form larger molecules, by the process we have called polymerization.

If more than one kind of molecule is involved in the joining-up process, the chemists call it co-polymerization. For most of these reactions we will simply call the big molecules polymers for the sake of simplicity.

CATALYST

One more word that is a favorite of plastics chemists should be explained. This word is catalyst. Catalyst is any substance that will affect the progress of a chemical reaction, entering into the reaction itself. There are a great many materials that will act as a catalyst under certain conditions. Some of these catalysts work to speed up or start a reaction. Others will, by their mere presence, slow down or stop a reaction. In order not to get these two kinds of reaction confused, we now call materials that slow down or stop a reaction negative catalysts or, more often inhibitors. This leaves the field clear for catalyst to mean any material that starts or speeds up a reaction. In plastics manufacture we make a great deal of use of catalysts and inhibitors.

In any discussion of plastics much is said about heat and temperature. We have previously mentioned the effect of heat or high temperatures on all organic compounds. This effect, together with the effects of catalyzers, inhibitors, and some other side line items, all have a part in helping us govern the polymerization reactions that are so essential in the formation of modern plastic materials.

We should not lose sight of the fact that the word plastics, as we use it, covers a large group of highly different materials. Different types of plastics are as different as different metals. No one would expect copper and tool steel to have the same characteristics of hardness or electrical conductivity. In the same way we must expect different characteristics and behavior from different plastics.

CLASSIFYING PLASTICS

Because there are so many different plastics it is difficult to classify them, or even to speak of them in general terms. There are many ways to classify plastics, such as listing them in the order of their discovery, or by the dates when they first appeared on the market. We might list them according to the basic materials that go into their make-up, and we shall try to do this in a limited way in the next chapter.

One of the simplest ways of classifying plastics is by their reaction to heat. This gives us a fairly workable method of separating the many kinds of plastics into two basic groups. Judging plastics by what happens to them, when they are heated, we can call some of them Thermosetting and another group of them Thermoplastic.

THERMOSETTING PLASTICS

The Thermosetting plastics will soften only once under the influence of heat. While they are hot, and in this softened condition, a chemical change takes place in the arrangement and bonding of the molecules so that the materials set or become hard. After this set takes place, the only effect of heat is to burn or char them. This charring or decomposition from heat begins somewhere in the range of 250 to 400 degrees F. There are a few exceptions to this rule as will be noted later.

THERMOPLASTIC MATERIALS

Thermoplastic materials soften or become "plastic" when heated, but no chemical change takes place in them when hot. When they cool, they again become hard and will assume any shape they have been molded into while soft. A thermoplastic material most of us are acquainted with is ordinary paraffin wax. While this is not considered one of our plastics its behavior under heating and cooling is an excellent illustration of what we mean by the term thermoplastic. Many of the thermoplastic types of plastics will not melt to a liquid--they just get soft when heated. Others can be melted until they become liquid and can be poured into a mold.

ACRYLIC PLASTICS

There are about fourteen basic kinds of thermoplastics. One type is the group of materials known as "Acrylics." This book deals largely with projects in which acrylic plastic material is used. Much more will be said about acrylics in the following chapters of this book.

In general, the thermoplastics have softening points ranging from 140 to 300 degrees F. In some cases the softening point is sharp-- that is, the material stays hard and non-plastic all the way up to a given high temperature--then suddenly softens. Other materials begin to soften at comparatively low temperatures, and get softer and softer as they get hotter. If we will remember that water at about 140 degrees F. can be tolerated by the human skin, and that water boils at 212 degrees F. we can estimate fairly close to the temperatures we have mentioned above.

In this section we have discussed some of the general facts about plastics and laid a foundation for some sort of understanding of the behavior of the atoms and molecules that go into their make-up.

TYPES OF PLASTICS IN COMMON USE

There are many general types of plastics being used today, and each general type has a number of variations. It is impossible to discuss in this book all of the types and their variations.

There are, however, a number of broad classifications of plastics that will include most of the materials on the market today. We have already mentioned the two broad classes of Thermosetting and Thermoplastic materials. In the following list some of the more common subdivisions under these classes are shown, listed according to the chief ingredient in the plastic material.

In the following brief discussion of these materials we shall try to point out any special uses or qualifications the materials may have, and to indicate the chief sources of the original materials that go into their make-up.

THE PHENOLICS

The word Phenolics is used as a general name for a class of thermosetting resins that include the original synthetic resin Baekeland first made by the reaction between phenol and formaldehyde. Phenol has the

THERMOSETTING RESINS	THERMOPLASTIC RESINS
Phenolics	Acrylics
Ureas	Styrenes
Melamines	Polyethylene
Polyesters	Polyamides
Epoxies	Vinyls
Silicones	Vinylidene Chlorides

In addition to these two lists of synthetic resins shown above, there are two other groups that are of importance or interest in any general discussion of plastics. These are:

<u>THE CELLULOSICS</u> (Made by modifying the cellulose molecule)

Cellulose Acetate	Cellulose Propionate
Cellulose Butyrate	Cellulose Nitrate
Cellulose Acetate-Butyrate	Ethyl Cellulose

<u>NATURAL RESINS</u>

Casein	Lignin
Shellac	

six-sided ring of benzene with a -O-H group attached to one of the angles of the hexagonal ring in place of a "H" (hydrogen). The H in this group is a flighty atom, easily led away from the parent molecule, and the presence of the -OH group also makes some of the other hydrogens at the points of the hexagon a bit nervous, hence the molecule is quite reactive, especially if it is heated a bit above room temperatures.

The formaldehyde molecule is also a jittery sort of chemical combination and when brought in contact with heated phenol it breaks up and joins onto the phenol in such a way as to allow for an almost endless linking up

Types of Plastics

of phenols and formaldehydes in long, rangy molecules, if the conditions of the reaction are properly controlled. This first product is liquid, or at most semi-solid. On further heating this material to just the proper temperatures and under the proper conditions, this first compound becomes an insoluible solid. This solid product is then ground up into powder or flakes. It is now the molding powder from which the final Bakelite plastic is made.

The molding powder is then mixed with a filler (wood flour, cotton fibers, chopped-up canvas, or perhaps mineral matter such as mica or talc) and the mixture is placed in molds under heat and pressure. The product is molded Bakelite resin. During this final or cure stage, chemical changes take place within the resinous material. This is thought to be a process of forming cross links between the long phenol-formaldehyde molecules that are the product of the earlier reactions. This cured, cross-linked material can never again be softened by heating.

If the resinous material, before the final heat-treating, is poured into molds without using any filler, and then heat-treated under pressure, beautiful "cast" phenolic resin substances are produced. These may be made in a wide range of colors, and may be highly polished. The result is an article of durable beauty. There are many uses for cast phenolics. Phenolic resins are used in many industrial materials, such as adhesives for holding the layers of plywood together, for making surfacing films for plywood or other material, and many other purposes.

Furfural is another member of the formaldehyde family. It is made from waste farm products such as oat hulls, cottonseed hulls, and the like. It also reacts with phenol to give resinous compounds similar to the phenol-formaldehyde resins. There are also other members of the phenol family, as well as other aldehydes, that form resins. The list of possible formulations is almost endless. We call all resins of this group "Phenolics."

In the case of molded phenolics, the filler usually amounts to about half of the bulk of the finished product. The nature of the filler used will have considerable bearing on the properties of the final molded product including its toughness, its ability to withstand shattering blows, and its ability to withstand heat.

Another large class of phenolic resin materials consists of the laminated products. If sheets of cloth, or of paper, or even of matted fibers, are spread with the liquid form of the resin, and then stacked layer upon layer, or perhaps wound around a suitably-shaped mandrel, and finally heat-cured, the resulting materials are called laminates. These find large-volume use in making a great variety of manufactured articles. Plywood, using a phenolic type resin as the bonding glue, is almost completely water-proof. It is also light and strong for its weight, and may be used as a building material in locations exposed to moisture or weathering.

Phenolic-base materials are largely resistant to the action of water, organic solvents, acids, alkalis, and many other substances. They do not burn readily, and extinguish themselves if no more heat is applied. These materials are useful up to temperatures above the boiling point of water (from 260 F. to 340 F., depending on the filler, the thickness of the material, etc.). Phenolics have excellent properties as electrical insulators.

Phenol originally was recovered from coal tar. Today, much of our phenol is made by synthetic processes from crude oil (Petroleum). Formaldehyde may be made from methyl (wood) alcohol and air. Modern methods of producing methanol consist of adding hydrogen to carbon monoxide under high pressures and temperatures.

THE UREAS

Previously we mentioned that Urea was the first organic compound to be made synthetically in the laboratory. This substance contains atoms of carbon, hydrogen, and oxygen in the molecule, along with nitrogen (chemical symbol N--nitrogen makes up about four-fifths of the air in the atmosphere).

When urea is mixed with formaldehyde under the proper conditions, a joining-up of molecules takes place much like that when phenol and formaldehyde are mixed. The resinous material coming out of this reaction is clear, or "water white." Because of this clearness, urea resin products may be made that are translucent (let light through) or that show a noticeable depth of surface. This gives a very pleasing appearance to these resins.

Urea resins are similar in their thermosetting properties to the phenolics. They may be molded in the same type of molds. They can be made in colors ranging from clear through translucent, and even white and black.

Molded urea resin products are not flammable, but will char at about 390 degrees F. and so are not suitable for making ash trays or other products that may encounter high temperatures. Urea resin articles show no temperature effects between 70 degrees F. below zero and 175 degrees F. above zero. Two trade names designating urea resin products are Bectle and Plaskon.

The intermediate resin product--the material before it is mixed with filler, and molded under heat and pressure--is water-soluble. This material is being used for treating papers ranging from paper towels to book-print stock. The water-soluble resin form is also used as an adhesive or glue in edge-gluing lumber cores, and other wood-gluing jobs. Urea resins are also used in making laminated sheets and stocks.

Urea resin plastics have outstanding electrical properties which are not much changed when the plastic is subjected to high humidity or exposure to moisture. Dry urea resins are remarkably resistant to corrosive fumes in the air. No known organic solvent will affect the moldings.

Urea may be readily synthesized from calcium cyanamide, which in turn is made from limestone, coke, and nitrogen from the air. The manufacture of the other ingredient of urea resins, formaldehyde, has already been discussed.

THE MELAMINES

The third member on our list of thermosetting plastics is the Melamines. The chemical substance called melamine is a carbon-hydrogen-nitrogen compound somewhat like urea, except that the melamine molecule is much larger and more complex than urea.

Melamine reacts with formaldehyde much as urea does. So far they have not been able to make "water white" melamines with the clarity of the urea compounds, but in other respects the melamine plastics are closely comparable to the urea plastics.

Melamine plastic products are sold under such trade names as Fiberite, Melmac, Plaskon, Melantine, and Resimene.

Melamine plastics are made using various kinds of fillers, depending on the properties most wanted in the finished product. The melamine-formaldehyde plastics are rigid, with a hard and durable surface that can stand up under a lot of rough handling. Most organic solvents, greases and oils, and many weak acids and alkalis have no effect on the molded part. The melamines have very low water absorption. Food containers made of melamine will not give any color, taste, or odor to foodstuffs stored in them. Melamines show an unusually low bacterial count when used as food containers or dishes.

Molded melamines are not sensitive to heat, and will withstand service temperatures in the range from 210 to 250 degrees F. These materials have excellent electrical insulating properties and may be used in most insulating applications.

If moldings are made using a high-grade cellulose filler, they can be made in an almost unlimited number of colors, and with considerable translucence and depth of color. The high-grade cellulose filler is made from cotton linters or wood pulp, purified by any of the standard chemical processes. Melamine moldings do not change dimensions under heat and cold, and do not absorb any appreciable amount of water.

If the intermediate resin material is combined with urea-formaldehyde intermediate, the resulting mixture makes an excellent glue that will withstand boiling water. Melamine intermediate is used to improve the wet strength of paper, to make cotton and woolen goods unshrinkable, in the tanning of leather,

Types of Plastics

and in many paints and other types of surface coatings.

At present, large amounts of melamine with high-grade cellulose filler are being used for molding heavy-duty dishware. These are designed for use in restaurants, hospitals, hotels, etc. Such dishes are practically unbreakable, can stand up under all known soaps and dishwashing compounds, and can take the wear and tear of dishwashing machines as well as careless handling by waiters and other personnel. Melamine dishware also has the advantage of being only about one-third as heavy as similar China ware.

The raw material, melamine, can be prepared synthetically from the fertilizer material, calcium cyanamide, and like urea from the same source, is in about the same cost range. The other component of melamine resins is formaldehyde.

THE POLYESTERS

In the process of heat-hardening the phenolic resins and the others we have been discussing, the chemical process that results in the combining of the smaller molecules into the big molecules of the polymerized resin, causes gaseous products to be given off. These may be water vapor, alcohol vapor, or several others. Because these gaseous products are given off during the curing process, the plastics must be cured under pressure in molds. If this is not done the gases can cause bubbles in the finished product.

The polyester resins do not have gaseous products given off during the curing, and therefore they may be manufactured into articles without the use of the pressure molds. For ordinary commercial and shop use, the polyesters are usually supplied in the form of a syrupy liquid. This liquid is actually an "alkyd" (more about this later) resin mixed with some reactive monomer, such as styrene. This monomer will, under the proper conditions, cause the alkyd molecules to bond together. The relatively small alkyd molecules, together with the linking monomer, form large molecules of the polymerized plastic resin. Sometimes other monomers with combing properties similar to those of styrene are used.

The alkyd resin mentioned in the preceding paragraph is originally made by causing an organic acid, with two acid groups on the molecule, to lose the elements of a molecule of water thus forming the "anhydride" of the acid. This anhydride is quite reactive and will go together with one of the unsaturated organic alcohols to form a large polymerized molecule. These resins are named from alcohol-acid---for some unknown reason they changed the "c" to a "k" and the "i" to a "y" and formed the word alkyd. Alkyd resins are not much used by themselves, but find large volume uses in making the polyesters, and in making the basic materials of a lot of paints, varnishes, and other surface-covering materials.

A mixture of an alkyd resin and a monomer will tend to polymerize slowly, even in a warm room, and much more rapidly if heated. To get really speedy action a catalyzer is added to speed up the reaction. The catalyzers are usually members of the families of chemicals known as peroxides or the hydroperoxides. These catalysts contain an unusually large amount of oxygen which is in a very reactive form.

If only the catalyzer is used, temperatures in the range of 180 to 240 degrees F. must be applied for a rapid reaction. If an accelerator such as benzoyl peroxide, is added, the reaction will readily begin at ordinary room temperature. Once the reaction is started, enough heat will be generated by the reaction itself to cure the polymerized resins without heat from any outside source. Statements that polyester resins cure at room temperatures are not quite accurate, because actually the chemicals in their reaction produce heat which causes the resin to set-up or cure.

One of the most common uses of the polyesters is in building up laminates using glass fiber cloth or matted glass fibers in sheets. These sheets of glass cloth or fibers are thoroughly soaked with the polyester resin mixture to which the catalyst and accelerator have been added. Since this mixture begins to set up just as soon as it is mixed, it must be applied to the work very soon after mixing. The mixtures usually have a "pot life" of about 15-30 minutes. This means that after that time the plastic sets up in the pot or mixing vessel and can no longer be spread on the work.

Polyester resins reinforced with glass fiber are strong and lightweight. The material can be sawed and sanded. The original resin surface is smooth and glossy.

The chief materials used in making polyester resins are the anhydride of phthalic acid and ethylene glycol. This latter chemical is a material of the alcohol family and is the same chemical material obtainable on the market as an anti-freeze solution for automobile radiators under the name of Prestone. The styrene monomer--the material most commonly found in commercial polyester preparations--is synthesized from petroleum. All these materials are today being made in large quantities and are relatively inexpensive. Also the glass cloth used for reinforcing is comparatively low in price.

EPOXY RESINS

The epoxy resins are comparative newcomers in the plastics field. As this is being written they have only been available on the market for a very short time, and as of now, their uses have not been as thoroughly studied as some of the older resins.

The term epoxy is the chemist's lingo which indicates that two groups of atoms have joined together by mutually clinging to an atom of oxygen. The epoxy resins are very similar in handling and in properties to the polyesters. They do have the ability to cling much more tightly to metal and other non-plastic surfaces. They make excellent adhesives for a great many uses. Epoxies, reinforced by glass fiber cloth may be used to patch holes in metal tanks--or to build up a hole or a deeply dented place in the sheet metal part of an automobile. Further study of these resins will be interesting.

THE SILICONES

The silicones are a very special and highly different class of thermosetting resinous materials. They are made by substituting atoms of silicon in place of carbon in certain complex organic compounds. If we remember that ordinary sand ("quartz" crystals) is made up of molecules of silicon dioxide, a material that is almost unaffected by any ordinary heat, weathering, or chemicals; and if we realize that the silicon atom in the midst of an organic molecule imparts some of this chemical sluggishness to the compound it is in, we can appreciate some of the differences that exist between ordinary organic polymers, and polymers that contain quite a bit of silicon in their make-up.

At present the silicones are rather expensive, and are limited to certain high-temperature applications. Because the atom of silicon is by nature rather unreactive, it is difficult to get it to enter into these compounds. The process requires a lot of "going around the long way." For this reason, silicone products are costly to produce, and expensive to buy.

This is another highly interesting class of plastic-like materials that would be a lot of fun to study. They range all the way from oil-like fluids to rubber-like solids and even harder materials.

THERMOPLASTIC RESINS

THE ACRYLICS

The most prominent member of the thermoplastic family as far as using plastics in craftwork is concerned is the group known as the Acrylics. The projects given in a later portion of this book deal largely with the Acrylics and methods of making various projects from them.

The acrylic resins are polymers of a chemical substance called methyl methacrylate. This is a rather complicated molecule that (as you can see from the name) is considered a derivative of acrylic acid. Acrylic acid also is called some other names by chemists. The main thing to remember about acrylic acid and it's derivatives is that they all have a double-bonded or unsaturated linkage.

Types of Plastics

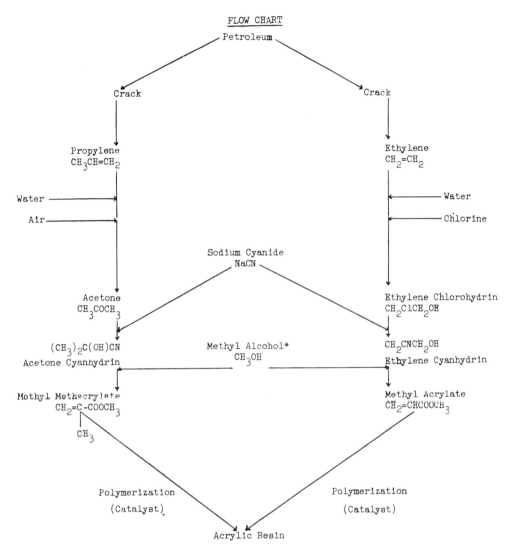

Fig. 1. Method by which methyl methacrylate resin may be prepared.

Fig. 1 shows a chart which gives one method, and the most common one, by which methyl methacrylate resin can be prepared. Let us point out that the chief starting chemicals are synthesized from petroleum, and that sodium cyanide, mentioned in the chart, is an extremely poisonous solid that can be made by heating calcium cyanamide, coke, and ordinary table salt in an electric furnace. Methyl alcohol (shown in the chart) has already been discussed.

As can be seen from this chart, many combinations of acrylic derivatives are possible. These, when polymerized, will give a whole series of resins varying from syrupy liquids to hard solids. The basic characteristics of all the acrylic resins are about the same, but we shall restrict our discussion to the hard, clear resin usually sold under the trade name of Plexiglas (Plexiglas is a registered trademark of the Rohm & Haas Co.) or Lucite (Lucite is a registered trademark of E. I. Dupont de Nemours Co.) The difference between the two trade-named products is almost impossible to detect, even by experienced chemists.

We are indebted to the Rohm and Haas Company for much of the information in this book dealing with acrylic plastics and methods of working with them.

The acrylic plastics lend themselves to many uses. They are thermoplastic, crystal clear, and will "pipe" light around corners. In the clear, transparent form, acrylic resins are as transparent as the finest optical glasses. The plastic may be produced in almost any color desired, and either the clear or the colored material is attractive and

Fig. 2. Fabricating Plexiglas display rack at Duke Bontz Co., Washington, D.C.

Types of Plastics

brilliant in appearance. The first impression on seeing a piece of transparent, clear acrylic is that it is extremely clear. Even in thick pieces or in a stack of thin sheets, the clarity is startling.

In the shop work section of this book the terms Plexiglas and acrylic resin are interchangeable.

The shop worker will have at hand for his working material, "cast" forms of this plastic. Acrylics may also be obtained in the form of molding powders, but using acrylics in this form in school and home shops is impractical. Cast shapes include sheets of various thicknesses, and rods and tubes of differing sizes and thicknesses. The material as obtained may have polished surfaces or the surface may be textured. As the term implies, "cast" materials are formed in molds during manufacture, and are usually cast between bonding surfaces of polished plate glass. This gives it a perfectly smooth, polished surface. See Fig. 2.

Because the material is easily formed, has outstanding optical properties, and because it is weather-resistant and relatively shatter-proof, much acrylic plastic sheet was used during World War II for airplane windows and canopies and also for periscopes in tanks, bomber "blisters," and other uses where clear vision was essential. Typical present-day uses include optical lenses, airplane windows, dials, instrument panels on planes, combs, costume jewelry, backs for brushes and for other toilet articles, powder boxes, and many other things of similar nature. Because of its light "piping" properties it is much used in surgical appliances, since almost the full amount of light admitted to one end of a rod (either straight or curved) is emitted from the other end. The light at the point of emission is "cold" and not hot as it would be if the light bulb itself were near the point of lighting.

Plexiglas sheet, rod and tubes may be sawed, drilled, machined, sanded, scraped, turned in a lathe, buffed and polished. Surfaces that have been rough cut may be polished to optical perfection. The material may be worked with the simplest of hand tools (it may even be planed with an ordinary woodworking plane, if necessary). Power tools are better for use in working acrylics because they allow for faster and more accurate work, and their working speeds can be adjusted to constant values.

When heated to around 240-300 degrees F., Plexiglas sheet becomes soft and pliable, and can be formed into almost any shape. It keeps the formed shape on cooling.

LIGHTING CHARACTERISTICS OF ACRYLICS

The light-transmitting properties--that is, the transparency--of solid acrylic plastic are of a very high order. Plexiglas sheet one-fourth inch thick will transmit about 92% of all the light reaching the first polished surface. The clearness of Plexiglas sheet is greater than that of most grades of glass and is very close to that of the very best "optical" grade of glass used for making lenses, etc.

One interesting application of acrylic materials is that of piping light around corners or to concentrate it on one particular spot. High-grade optical glass has this same property. But because of the cost of such glass, it's heaviness, and it's hardness, it is not used much for such purposes. Acrylics, are light in weight, easily formed into any desired shape because of their thermoplastic qualities; and the rough surfaces are easily given a high polish.

A short, non-technical discussion of the physical principles behind this phenomenon may be of interest. Whenever light passes from one transparent substance into another of different density, the light beams tend to change direction at the interface--at the surface where these two substances join. This bending of the path of light ray is known as refraction. The amount of bending depends on the difference in density between the two transparent materials that meet at the common surfaces, and it is also dependent to some extent on the molecular composition of the two different transparent substances.

Primarily we must know that when light travels through the length of a transparent rod that has a highly polished surface in contact with the air around it--(or when it travels into the edge of a sheet with polished faces) the light inside the transparent material is sent back from the polished surfaces of the material--back and forth with great efficiency--until it comes out the other end or side of the material. That is what happens when light is "piped" through the length of a rod or the edge of a sheet.

This effect is largely due to the degree of bending (or the angle of refraction) as light passes from plastic to air, or from air to plastic. If light inside a sheet of acrylic strikes the surface at any angle with the perpendicular greater than 42.2 degrees, it is bent so that it cannot escape into the air, but must be reflected back to the opposite face of the plastic material. If these two faces are parallel, then the light keeps bouncing back and forth until it comes out at the far end. In "edge lighting" of sheet acrylic, or in lighting through the end of a rod, this batting of the light rays back and forth from the inside of the plastic-air surface causes most of it to be finally carried through to the first surface where the light will hit with a greater angle than the 42.2-degree angle (this is called the "critical" angle of the air-plastic interface). This means that light will travel through to the end of the rod or the far edge of the sheet.

The reason that acrylics are so good at this light piping is primarily because they are so clear that most of the light finally gets through the material without being absorbed or "blotted out." In ordinary glass, small amounts of colored materials left in the glass will absorb much of the light if it has to travel more than a few inches in the material. Acrylics transmit a large portion of the total light that enters the rod or sheet of the plastic material. If the plastic is bent, so long as the angle of bending is less than this critical angle, the light continues to travel through the length of the plastic until it reaches the opposite side or surface. For rough measurement purposes, we can say that if the radius of the bend in acrylics is more than three times the thickness of the plastic (that is the distance between the reflective sides) the light will pass satisfactorily around the curve.

In piping light through a piece of polished acrylic rod, there are two noticeable effects. One is that the light appears to be more brilliant at the end of the rod, than the surrounding light coming from the same source. This is because practically all the light which enters the rod near the source of light is transmitted, without spreading out, to the far end of the rod. Light given off from the same source into the air is directed outward from the center of illumination in all directions and becomes more diffused as it travels outward from its source, so that the illumination falling on a given area decreases with the "square of the distance" from the light. In the acrylic rod, none of the light entering the light-end of the rod is diffused, but comes out the end of the rod having lost only that amount of light that was absorbed during passage through the acrylic. The second noticeable effect is that the light coming from the end of the rod appears much "cooler" than the same amount of light shown on the receiving surface directly from the light source. This is because much of the penetrating heat rays of the light have been filtered out in the plastic. This latter is of great service in surgical instruments designed to deliver light through an acrylic rod to a point in the interior of the body. The light from the lamp is delivered to the point where it is wanted--the heat is filtered out.

CRITICAL ANGLE OF REFLECTION

Another thing to remember in dealing with lighting problems in acrylic plastics is that this critical angle of reflection or refraction (which ever way you want to look at it) is true only when air and plastic come together. If some material, such as a tightly adhering paint, is applied to the surface of the plastic, the angle of bending is changed and that painted surface reflects light at a different angle, so that it strikes the opposite face of the plastic at an angle that will let it go through the air-plastic surface and become visible. In this way, a design painted on the far side of a sheet of plastic will become luminous from the light being piped through the plastic sheet.

Other methods of disturbing this plastic-air surface also produce luminosity which is visible from the opposite of the sheet. Sandblasting, or any other roughening of the plastic surface, causes light to be reflected back at a large enough angle to be visible from the opposite side of the sheet. This can also be done by grooving or otherwise disturbing the angle of the air-plastic interface, as will be demonstrated in the projects.

POLYSTYRENE

The polystyrene plastics (usually called "styrenes" for short, are among the most versatile of the plastic resins. (Versatile means that they can be used in many different ways. These plastics may be compounded in so many different combinations that their properties and variations are almost endless. Styrene may be polymerized alone or co-polymerized with many other materials. The styrenes are thermoplastic.

The first styrene resin was noted in the laboratory more than 100 years ago. But for many years the reaction was only a laboratory curiosity. The introduction of styrene polymer into the commercial field had to wait for an economical method of preparing it in large quantities. This material is made up of an unsaturated ethylene molecule in which one of the four hydrogens has been replaced by the benzene ring. Ethylene looks like this: $H-\underset{H}{C}=\underset{H}{C}-H$.

Styrene looks like this: $H-\underset{H}{C}=\underset{H}{C}-(C_6H_5)$.

Evidence points to the probability that styrene polymerizes by joining up of the ethylene part of the molecule at the point of the double bond, making long-chain molecules. These long-chain molecules without any cross-linking, are typical of the thermoplastics. In the thermosetting materials, the heat and pressure causes cross-links to form, and these cross-links prevent the molecules from moving around--that is, prevent the material from softening, when heat is applied.

The styrene polymers are very stable from a dimensional standpoint, and keep their physical properties (hardness, etc.) over a fairly large temperature range. Fungus will not grow on these plastics, and hence they are useful in the steaming tropics. Styrenes have excellent electrical properties plus low water absorption. This makes them useful in many electronic applications, such as television insulation.

Because they are inert toward many chemicals, styrene plastics may be profitably used for trays to hold photographic chemicals, or for laboratory apparatus that must resist the action of acids, alkalis, and other corrosive chemicals. Films of styrene polymer show very little tendency to allow water vapor to pass through, and so are useful as wrappings to keep moisture in or out.

Styrenes are not very resistant to strong light. They are more brittle than some other plastics, and hence more breakable. Styrenes have a relatively low softening point (about 190 deg. F.), but in the range from -40 to +160 degrees their physical properties are practically constant.

Because styrene monomer can co-polymerize with a wide range of other plastic-forming materials, special formulations can be prepared that will improve one or more of the properties of the plain styrene plastic. Usually, this improvement is obtained at the expense of some other property.

There are four general formulations of styrene plastics being marketed for molding purposes. These are: (1) General purpose, (2) High heat resistant, (3) Medium impact, and (4) High impact. These different formulations favor some particular property of the finished product, as the name implies. For example, while the heat-distortion point of general purpose (1) is about 180 deg. F., the high heat resistant (2) material does not soften below 230 deg. F. On the other hand the high impact (4) formula gives a plastic softening at an even lower temperature than the general purpose material, but is much more resistant to a sudden blow with a hard object.

Styrenes are on the market under such names as: Ampacet, Koppers, Catalin, Lustrex, Bakelite Styrene, and Styron--just to mention a few. There is a perfect maze of trade-name materials made by combining styrenes with other plastics to bring out certain basic properties. These materials are being used to make housings and containers formerly made by metals, wash basins, pipe, boat decks, and a host of other items including large household and industrial items that require toughness, dimensional stability, resistance to chemicals, and easy formation or moldability.

One special styrene polymer product that will be used in a project given in this book is an expanded or bubble-filled form sold under the name of Styrofoam (Styrofoam is a registered trademark of the Dow Chemical Co.) This material is a plastic foam that has been set to form a solid material. Because it is largely air-filled, the foam is very light in weight for a given bulk. One grade on the market only weighs about one and one-third pounds per cubic foot.

The air-bubbles or open cells in this foam are sealed off and are independent of each other so the material will not soak up water, and a piece of it on water will float indefinitely. It does not transmit heat, and hence makes an excellent insulating material for such applications as insulating refrigerators. It is also used in a sandwich type of construction in which "boards" of the fluffy plastic are covered on each side with thin sheets of wood or metal, or other material, giving a thick, lightweight, strong building material. Styrofoam is also widely used in making display and novelty items, such as Christmas decorations.

The material is easy to bond or cement. It is resistant to heat not exceeding 175 deg. F. over continuous periods of time. In water it will hold up or support about 55 pounds of load per cubic foot, and because it will not "waterlog," makes excellent buoyancy material for use in boats, life-preservers, etc.

Some styrene co-polymer mixtures have a rubber-like flexibility, and have been widely used in sheathing electric cables. These materials are in the lower-cost range for plastics, and for this reason have come into wider use than some of the more expensive types.

Although the polymerizing action of styrene has been known to chemists for more than a century, the reacton was little more than a laboratory curiosity until the recent advances in use of synthetic polymers. Styrene polymerizes very slowly at room temperatures. But at higher temperatures, and especially if oxygen is present, the polymerization reaction is quite rapid.

Today, styrene is synthesized in large quantities by combining ethylene (obtained from gases produced when petroleum is "cracked" by strong heat), together with benzene (which may also be obtained from petroleum by synthetic methods, as well as from the older source, coal tar). Ethylene plus benzene forms ethylbenzene. This product is then cracked by heating to give the unsaturated or double bonded ethylene benzene, or styrene.

POLYETHYLENE

Polyethylene plastic was first developed in England during World War II. Shortly afterward it was introduced into the United States. Because of its highly favorable electrical properties, polyethylene was first used as an insulating material in flexible co-axial cables (such as are used to transmit television signals, and other high-frequency electrical impulses).

Polyethylene is milky-white and translucent. It feels warm, smooth and waxy to the touch. In thick sections it is stiff and rigid. In thin sheets it is very flexible, and this flexibility has led to a growing use of this material in making "squeeze bottles" for dispensing all sorts of liquid and semi-liquid materials. The polyethylene bottles are more expensive than glass, but are unbreakable and "squeezable," so that their use is increasing very rapidly. Polyethylene softens at 220-240 deg. F. It will lose its shape if put under a load of 66 pounds per square inch and heated to about 122 F. for some time.

Polyethylene will absorb practically no water. It resists the action of all acids, alkalis, inorganic chemicals, etc., and there are no known solvents, either organic or

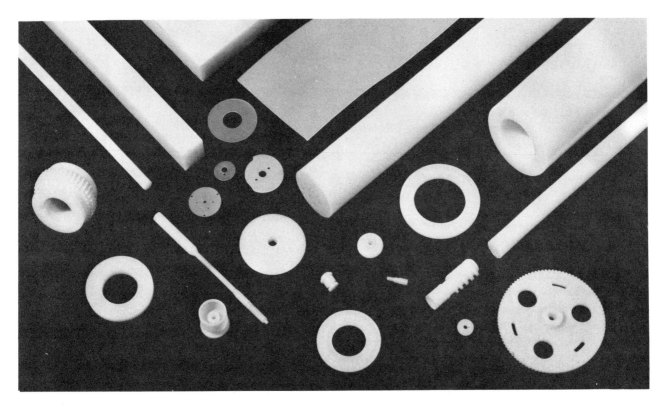

Fig. 3. Stock shape and parts fabricated from nylon.

inorganic, that will attack it at ordinary temperatures. At higher temperatures (around 140 F.) it dissolves readily in quite a number of the common organic solvents. Because it is so soft and resilient, it is excellent for packaging delicate or fragile articles such as thermometers and watches. Polyethylene thick film has largely replaced heavier and more expensive lead as a sheathing material for electrical cables (such as telephone cables). In total pounds of use, it is one of the fastest growing of all synthetic polymers.

The common method of polymerizing ethylene gas is to put it under tremendous pressures (around 15,000 to 30,000 pounds per square inch) and heat it to about 370 deg. F. The high pressure literally squeezes the unsaturated ethylene molecules so close together that they join up under the influence of heat to form the large polyethylene molecules. Some newer processes for making polyethylene make use of catalysts at near-normal temperatures and pressures. This newer type of polyethylene is stiffer, more heat-resistant, and has a higher tensile strength (resistance to being pulled apart) than the older form. Polyethylene pipe is being increasingly used for piping liquids or gases under high pressure. Such pipe may also be used for piping water in locations where there is danger of freezing. Ordinary metal pipes usually burst when water in them expands on freezing, but polyethylene pipe merely "gives" enough to take care of the extra expansion, and then returns to the normal size when the ice has melted.

The raw material for polyethylene is ethylene gas. This small, unsaturated molecule ($H-\underset{H}{C}=\underset{H}{C}-H$) is a by-product of cracking petroleum oils and can be obtained in large quantities from this source.

POLYAMIDES

Polyamides are plastic materials made up of polymerized polybasic organic acids and polyfunctional amines. To put this in simple language, the two kinds of molecules making up the polyamides are (1) organic acids that have more than one acid-acting group attached to the molecule; plus (2) Nitrogen-hydrogen-carbon compounds that have more than one

amino group (H-N-) on the molecule. These
             H
amines are very similar in chemical construction to urea and melemine.

The original commercial polyamide was developed by the DuPont Company under the trade name "nylon." This brand name has now become a household word, spelled with a small "n," and the term "nylons" refers to this family of polyamide plastics. DuPont now calls their polyamide materials Zytel (Zytel is a registered trademark of E. I. DuPont de Nemours Co.) They have licensed a number of other manufacturers to make nylons under their own trade names.

Most of us remember that the original use of nylon was in the form of fine threads to make women's stockings, and was marketed as a competitor for silk used for the same purpose. Quite a number of these polymerized materials, called plastics in this book, can be squeezed through tiny holes to produce threads that can be knitted or woven into cloth or synthetic fabric. Because of space limitations such usage of plastics will not be covered in this book. When we speak of "nylon" in this text, we refer to the material that has been molded or extruded into rods or sheets, and not the threads that go into making stockings. See Fig. 3.

Nylon plastic is light in weight, weighing slightly more than a similar volume of water. Its burning rate is extremely slow, and it is self-extinguishing--that is, it will go out of its own accord as soon as the source of heat that originally started the fire is removed.

Molded nylon parts are tough and durable, very strong, and highly resistant to wear. They also resist corrosion, and are good heat insulators. Nylon parts show no warping due to heat at temperatures under 250 deg. F. This material is difficult to mold, because it is not very fluid at molding temperatures, and it is necessary to control the temperature very closely during the operation.

Thin-section moldings of nylon may be used for making small gears, cams, and other machine parts. It is satisfactory for such uses because it is tough, and because it is self-lubricating. In many instances these parts can be molded from nylon at less expense than machining them from metal. Nylon's resistance to most chemicals, except mineral acids and phenols, is outstanding.

From the chemist's viewpoint, the long-chain molecules of the polyamides are very similar to the long-chain molecules of natural proteins- chemical substances that make up the bulk of most meat, and animal fibers such as silk and wool.

Aside from its use in thread form in making fabrics, most nylon is used for molding small machine parts. One of the chief factors in making such machine parts of value is that these parts may run without using lubricating oils. In food processing machinery, where the lubricant might adulterate the product or introduce objectional tastes or odors, nylon machine parts proved invaluable.

Also in the textile industry, where the danger of lubricating oil or grease spoiling or staining the delicate fabrics being woven, or otherwise produced, the no-lubrication feature of nylon parts is highly valuable.

In many applications nylon machine parts will outwear similar parts made of metal. Nylon can be formed by most of the ordinary metal-forming technique to tolerances roughly equivalent to those required of the corresponding metal part.

Another factor in nylon machine parts is their resiliency or "give." This makes for quieter running, with less vibration and noise. Uneven strains on different parts of a nylon gear or bearing are equalized by this resiliency, and there is not the danger of fracture of the part when subjected to unequal stresses as there is with metal parts.

The raw materials that go into nylon are more expensive than some of the others we have mentioned. The statement made about nylon when it first came on the market--that it was made from coal, air and water--is very much of an over-simplification. As a matter of fact, the polybasic acids and the polyfunctional amines that go into making nylon may be synthesized from very simple products, but these synthetic processes are round-about and difficult. This is one of the reasons nylon products are still in the higher priced class of polymerized synthetic resins.

Types of Plastics

THE VINYL PLASTICS

The vinyl "radical" (a group of atoms acting as a unit) is the same as ethylene, but with one of the hydrogens missing and ready to be replaced by another atom or linkage to some other group of atoms. For example, vinyl chloride has this structure: $H-\underset{H}{C}=\underset{Cl}{C}-H$.
Here we see again the two double-bonded carbons. The naming of this group of atoms goes back to the original confusion of the organic chemicals when they were naming new compounds. The group represented above is the "vinyl" group, if the chlorine atom (Cl) is omitted.

All the polyvinyl plastics start out with a substance that has this unsaturated vinyl group to which some one atom or group of atoms has been added in the position of the chloride given in the foregoing formula. Vinyl alcohol, vinyl acetate, and vinyl bromide, as well as vinyl chloride are frequently used in polymerization reactions. In fact, styrene could very well be called vinyl benzene. The styrene plastics have been covered under another heading.

The polyvinyl plastics were originally introduced as rubber substitutes. These materials are typical polymer resins, but they may be given rubber-like qualities by adding suitable plasticizers. A plasticizer, when added to any material, does just exactly what the word seems to mean--that is, it makes the material more plastic--or softer and easier to mold into shape. One way of thinking of a plasticizer is that it is a solvent which does not evaporate. One example would be adding water to clay (if the water would not dry out later), or adding corn syrup to fudge to keep it from getting hard and grainy.

The most common plasticizer for vinyl resins is tri-cresyl phosphate (the TCP of gasoline advertising fame). We will not give the complicated formula for TCP here-- enough to say that when it is added to vinyl resins it makes them pliable, stretchable and rubber-like. One trade-marked product, Koroseal, contains about 40% TCP in its formula. There are so many vinyl polymers and co-polymers on the market that it would be impossible in this limited space to list all the trade names covering these materials.

The raw material for vinyl plastics is ethylene gas--the same starting material we found in the manufacture of polyethylene. It comes from cracking petroleum, and there is plenty of it.

POLYVINYLIDENE CHLORIDE PLASTICS

The "vinylidene" group or radical is one in which both the hydrogens from one carbon of ethylene have been replaced by some other atom or group of atoms. In vinylidene chloride the molecule has a structure something like this: $H-\underset{H}{C}=\underset{Cl}{C}-Cl$.
Here, two chlorine atoms are hooked onto one of the carbons in the unsaturated, double-bonded, ethylene structure.

Polymerized vinylidene chloride is a relative new-comer to the plastics family. Its chemical and physical properties are similar to those of the polyvinyl plastics. It is more easily dissolved in some of the common organic solvents, and it may be processed at a lower temperature than the vinyl compounds. Its chief uses are in coating fabrics, and in making films, and extruded and molded parts. A polyvinylidene film is marketed under the trade name Saran Wrap, and is gaining wide acceptance as a household film for wrapping foodstuffs and other materials.

Another trade name for polyvinylidene chloride materials, especially the molded or extruded type, is Geon.

Polyvinylidene chloride parts molded by injection molding are found as valve seats, acid dippers, filter parts, nozzle tips, and many others. Tubing of polyvinylidene chloride makes an excellent hose for piping ethyl gasoline and other petroleum products, since the hydrogen materials do not soften it like they do a natural rubber hose.

The starting material for this family of plastics is vinylidene chloride. This chloride is made from acetylene gas. Acetylene has a formula something like this: $H-C=C-H$.
The two carbons are bonded by a triple bond instead of a double bond as in ethylene. This triple bond is highly reactive thus making

polymerizing materials containing it a relatively easy job. Acetylene gas may be prepared by dripping water onto calcium carbide. For a long time lamps which generated acetylene in this manner were used by miners and campers, and the acetylene gas burns with a highly luminous flame when it comes out of a suitable nozzle. Acetylene is also used in large quantities in oxy-acetylene welding torches, because when it is premixed with oxygen it burns with one of the hottest flames man is able to produce. The calcium carbide for making acetylene is manufactured by heating limestone with coke in an electric furnace. It is easily and inexpensively made in this manner.

SYNTHETIC RUBBERS

Any chemist will shudder when he hears the term "synthetic rubber" used as we ordinarily use it. He would prefer to hear the word "elastomer." These materials are not like rubber, chemically, even though they do have many of the physical properties we usually associate with natural rubber products. Most of these elastomers (or synthetic rubbers) are plasticized synthetic resins that have been treated so that they are soft and elastic like rubber. Some of these elastomers have certain properties that make them better for specified uses than natural rubber.

POLYFLUORO HYDROCARBONS

The element Fluorine (symbol "F"), has a strong ability to hang onto other elements. For this reason, chemical compounds containing fluorine are usually very stable or hard to break up. If fluorine is substituted for some of the hydrogen atoms in a normal compound of carbon and hydrogen, we call these compounds Fluoro Hydrocarbons. When such a compound is polymerized, we get the polyfluoro hydrocarbons. Two of these are worth mentioning here, polytetrafluroethylene and polytriflurochloroethylene. Let us stop to explain these long names. Chemists always try to make the name of a compound tell you just how the material is made up. If you will examine each of the long words above a syllable at a time, you will see that the first one comes out like this: Poly-(many)-tetra (four)-fluoro (fluorine atom)-ethylene. The substance referred to has four fluorine atoms on an ethylene molecule, and these molecules have been polymerized into a large molecule. In the case of the other one, three fluorines and one chlorine are attached to the primary ethylene molecule.

These materials are on the market today. The one with four F atoms is called Teflon-- the one with one chlorine and three F's is called Kel-F, Fluorothene, or Bakelite CF-3. These materials are to a certain extent thermoplastic, but because of the extreme stability of the fluorine compounds, they must be heated a lot hotter than the other organic compounds we have mentioned before they will start to decompose under the influence of heat.

These polyfluoro plastics have several unusual characteristics. One is that they are smooth and waxy to the touch, and practically nothing will stick to them--they won't even stick to themselves, even under heat and pressure. No satisfactory glues or adhesives have yet been found for them. When molded into various types of parts, the resulting part is flexible, and will withstand much higher temperatures than any of the other plastics mentioned, when used as gaskets, O-rings, etc. Films of these materials are still quite flexible at minus 100 degrees F.

Fabrication of parts is difficult. Teflon, for example must be formed into the desired shape by heating the powdered resin under high temperature and pressure. Even then the material does not melt and flow, but sinters--that is the outer surfaces of the small particles become softened and the particles are welded together or sintered by the pressure being applied.

These polyfluoro resins are in the higher cost brackets because of the cost of chemical processes involved, and because of the difficulty of fabricating the finished pieces.

Types of Plastics

OTHER PLASTIC-LIKE MATERIALS

THE CELLULOSICS

Cellulose is the main constituent of the woody matter of plants and trees--cotton is almost pure cellulose. By treating purified cellulose with certain chemicals, the enormously large molecule of cellulose is so changed that materials are produced which are much like the polymerized plastic substances we have been discussing. Since these compounds are not first polymerized, but have already been polymerized in nature in the form of the large cellulose molecule, they are not classed with our synthetically polymerized resins.

These Cellulosics as they are called, are thermoplastic resin-like materials that have found wide use in industry. One or two of the projects in this book deal with cellulosic materials.

The natural cellulose molecule is a long-chain molecule having numerous "reactive" groups on it. If cellulose is treated with nitric acid (in the presence of sulfuric acid) a material called cellulose nitrate is produced. This modified cellulose or cellulose nitrate was first prepared about 100 years ago in an attempt to find a substitute for ivory in making billiard balls. It was later used in making molded articles and was marketed under the name of pyroxylin or celluloid.

One of the principal drawbacks of the cellulose nitrate materials is that they are dangerously inflammable. For many years clear films of cellulose nitrate were used as the backing for photographic film, both for ordinary photography and for the early movies. When they were using this nitrate film, the fire hazard in a movie projection booth was terrific, and many disastrous fires were started from this film. The inflammability of such film is almost explosive in its violence.

In an attempt to get away from this dangerous fire hazard, chemists developed a cellulose acetate film, made by reacting cellulose with acetic acid (the acid in ordinary vinegar). Cellulose acetate has a very slow burning rate (as compared with the nitrate product), and is still largely used in photographic film (know as "safety" film).

Cellulose acetate is normally thermoplastic. Because it can be easily plasticized it can be made in forms varying from stiff and rigid to a rubber-like material, and is used to a great extent as a molding compound. Cellulose acetate films--in addition to their photographic uses--are used in making adhesive films and recording tapes, and for molding electrical and heat insulating materials. Cellulose acetate, when properly plasticized and forced through small holes, makes a long thread-like filament that is woven into cloth, called acetate rayon. Cellulose acetate sheets varying in thickness from three-thousandths of an inch to one-fourth inch are readily available on the market, and are used for making a wide variety of articles.

In addition to nitric and acetic acid, other organic acid groups may be added to the cellulose molecule. Mixtures of two or more acid groups in this reaction produce materials having widely differing properties. One we are particularly interested in in this book is cellulose acetate-butyrate. This material contains the vinegar acid called acetic acid and also butyric acid, which is the evil-smelling acid formed in butter when it gets rancid. Cellulose acetate-butyrate has found many applications. It has excellent molding properties, it has high impact strength (can be hit with a hammer without breaking). It is very tough, and is dimensionally stable. It is not affected by dilute sulfuric or hydrochloric acid, and is resistant to the action of gasoline, oils, and other hydrocarbon materials. However, this material is readily attacked by many of the common organic solvents, and is also easily softened by heat. It is not suitable for carrying heavy loads.

Cellulose acetate-butyrate is much used in making articles that are likely to receive shattering blows. We are probably familiar with it as the yellow-transparent handles found on chisels, screwdrivers and other tools that are likely to be struck with a hammer. Plastic hammers are usually made with this material.

Cellulose acetate-butyrate is obtainable on the market under several trade names. Tennite II or Tennite CAB are names of this material and will be used in projects in this book. This plastic is one of the less expensive plastic materials.

A number of other cellulosics are being introduced in the market, and the wide-awake user of plastic materials should keep in touch with the development of these relatively low-cost materials.

The relatively pure cellulose used in making the cellulosics can be obtained from wood pulp such as is used in paper making. Another economical source is cotton linters, the very short cotton fibers that stick to the seeds when cotton is ginned. This raw material is therefore readily obtainable in larger quantities, and hence is not expensive. The organic acids used for combining with cellulose in making the cellulosics are also obtainable in large quantities and at reasonable prices.

NATURAL RESINS

In the first chapter we listed three natural resins. These are materials found in their reactions. A mention of them is necessary in any discussion of plastic materials.

SHELLAC

This is a gum resin deposited by insects on certain plants--largely found in Asia. It is thermoplastic, and is used with a suitable filler in molding many different kinds of objects. Formerly, it was the chief material used in making phonograph records. Its chief disadvantage is its low heat-resistance--it softens at about 175-180 degrees F. Alcoholic solutions of this material are used in surface coatings of wood and other materials.

CASEIN

Casein is the "curd" in milk. The curd is precipitated from milk, then mixed with filler material and ground in the form of a dry powder. Dyes dissolved in water are sprayed onto the powder which is then extruded from hydraulic presses, cured in formaldehyde and dried. The material is obtainable in the form of sheets, rods, and tubes.

LIGNIN

Lignin is the gummy binding material found in most woods which binds the cellulose fibers together. When separated from the cellulose, lignin is a thermoplastic resin. It is largely used as a re-binder of wood pulp. Masonite boards are made by grinding up wood waste, separating the lignin from the pulp, and distributing the pulp into "boards" of suitable thickness. The lignin is then added to the pulp and the boards are "set" under heat and pressure. Lignin reacts with amines, furfural and phenol to form thermosetting materials.

WORKING WITH ACRYLICS

Acrylic plastics because of their beauty, optical clarity, and the fact they can be easily worked, are well suited for use in making a great variety of outstanding craft projects.

In this section on Handling and Working with Acrylics, we will discuss Storing, Sawing, Machining, Polishing, Heat Forming, Cementing, Bonding, and Cleaning. Becoming familiar with these basic procedures is the first step toward turning out projects described in this book, and may well be a starting place for creating projects of your own design, in this colorful material.

Acrylic plastics are made by a number of concerns--the two best known being Plexiglas made by the Rohm and Haas Co., and Lucite, made by E. I. du Pont de Nemours & Co. Instructions given here apply to all of the acrylics. In this section, where the term "plastics" is used, acrylics are meant.

FORMS IN WHICH PLASTICS COME

Plastic stock as used in the school and home shop, comes in the form of sheets, rods and tubes. In sheet form it is available from plastic supply houses in thicknesses of 3/64 in. to 4 in. Plastic tubing may be obtained in outside diameters of 1/4 in. to 6-3/32 in. Round rods come in 1/8 in. to 3 in. diameters; and square rods 1/4 in. to 4 in. square. Many, but not all sizes are available in a wide selection of colors.

PLASTIC HAS POLISHED SURFACES

Since acrylic plastic is a thermoplastic material and softens with heat, and unsupported lengths of sheets, rods and tubes are subject to warping and bending, proper storage is of considerable importance.

Proper storage is also important, to protect the plastic surfaces from scratches. These materials in most sizes, come from the manufacturers with highly polished surfaces. Any scratches on the surface or any hazing of the surface, spoils the appearance of the finished project. The polished surfaces are relatively soft and easily scratched.

Plastic sheets usually come from the manufacturer covered on both sides with heavy masking paper. This paper which is coated with pressure-sensitive adhesive, clings tightly to the surface of the plastic. The masking paper should be left on the material as long as possible; through working operations such as sawing, drilling, planing of edges, etc. Be sure to keep in mind the fact that protection provided by the masking paper is limited, and that an object like a tack, or a chip of metal, can easily cut through the paper and make a bad scratch on the surface.

STORING PLASTIC STOCK

Plastic sheets may be stored flat, with one sheet upon another on a flat surface; or in a vertical position on edge. A still better way to store plastic sheets is at a slight angle. See Fig. 4. The storage bin shown here, is made of flat, well-supported sheets of plywood. The sides are inclined a few degrees from vertical--just enough so the sheets are stored flat against the side and are held in place by their own weight. When stacking plastic sheets for storage, both sides of each sheet should be brushed using a soft-bristle brush, or soft cloth, to remove particles of grit, etc. that may be clinging to the masking paper.

Store sheets of the same size and thickness together as much as possible.

If the masking paper has been removed from a sheet that must be returned to stock, cover the plastic with masking paper previously removed from the stock, if available, or wrap the sheet in soft tissue paper.

Plastic rods and tubing are not normally covered with masking paper, and must be handled carefully to avoid scratching of the polished surfaces. Rods and tubing should be stored flat in shallow bins. No ends should extend beyond the bin as unsupported stock will sag under its own weight.

The storage room should be well ventilated; humidity should be moderate. Hot, dry air will eventually dry the adhesive on masking paper, and make the paper hard to remove. Too much humidity will cause masking paper to deteriorate to the point where it no longer provides effective protection for the sheets.

Sunlight shining on stored plastic sheets

Fig. 4. Storing Plexiglas in upright, but slightly tilted storage bins.

will not affect the plastic, but it may cause adhesive on the masking paper to harden. Excessive heat in the storage area may cause crazing or other deterioration of the plastic surface. Do not store plastic near hot water or steam pipes.

A number of organic solvents will attack plastics. Even fumes in the air, such as might come from a paint-spray booth, may cause trouble.

Working with Acrylics

TRANSFERRING DESIGNS TO MATERIAL

Masking paper on plastic sheets may be used to good advantage in laying out designs-- lines indicating where to saw, drill holes, etc. The lines may be drawn right on the paper using a lead pencil. Designs may be traced onto the masking paper using carbon paper, or the designs may be drawn on separate sheets of paper and pasted on the plastic, using rubber cement.

In cases where it is necessary to mark on a plastic sheet from which the masking paper has been removed a China marking pencil may be used. The mark may be removed with a soft cloth. Or the markings may be scratched or scribed on a plastic surface using a scratch awl, or other sharp pointed instrument. It is not advisable to scribe along a line where a bend is to be made as even a slight cut will provide a starting point for a crack or break, particularly if internal stresses have been set up in the plastic while bending.

Masking paper MUST be removed before heating plastic for forming operations.

To remove masking paper, lift it at one corner, then pull away from the surface with a slow, steady pull. Do not try to jerk the paper off. Spots of adhesive still sticking to the polished surface may usually be removed by

Fig. 5. Sawing sheet plastic on circular saw with a fine-toothed blade. Guard has been lifted to show action of saw.

Fig. 6. Sawing large panel of corrugated Plexiglas on circular saw. Guard has been removed to show saw blade.

touching them with the sticky side of a piece of masking paper. Adhesive on hard-to-remove masking paper may be softened by heating the paper covered plastic sheet a short time--never more than a minute, in an oven heated to 250 deg. F.

SAWING PLASTICS

The acrylics may be worked with either woodworking or metalworking tools. These materials are harder than most woods and softer than most metals. On most jobs best results are obtained by using power-driven tools with metal-cutting blades.

For most projects, the pieces of plastic needed must be cut from stock sheets, rods, or tubes. The easiest way to cut plastic is to saw it. A power-driven circular saw is best for making straight cuts, Figures 5 and 6. For making curved cuts in thin sheets, a jig saw may be used. A band saw is best for making curved as well as straight cuts in thick stock.

If we remember that acrylic plastic is thermoplastic, we can be guided in our selection of cutting tools, cutting speeds, etc. When a saw or other tool runs fast enough to melt the plastic, the tool will gum up and a poor job of cutting will result. This is particularly true with the thicker materials. Plastic is a very poor conductor of heat, so heat generated by a cutting tool remains largely at the point of cutting, causing the plastic to soften.

Circular saw blades used in cutting plastics should be hollow ground (blade is thinner below teeth than teeth) or the teeth should have a slight set. Remember that plastic is expensive and that the chips cut away by the saw blade are waste. The thicker the saw and wider the cut, the more material is lost in each cut.

A sharp saw should be used at all times. A dull saw will tend to heat up the work, and may cause damage not only to the work but to the saw blade as well. For general purpose use in school and home workshops, a hollow-ground, 7-tooth per inch, 8-inch diameter metal cutting saw blade, run at the conventional speed with a 1750 r.p.m. electric motor, will be satisfactory. The saw teeth should be filed the same as for cutting soft metals such as brass or copper--the front angle of the teeth should be at right angles with the cut. The saw teeth should be side dressed by "rubbing" an oil stone very lightly against the revolving blade.

In all mechanical cutting or working of plastic, the tools should have only a very slight rake. This means that the cutting edge should strike the work at such an angle the action is a scraping action rather than a cutting action. Because of its toughness, more power is needed to saw acrylic plastic, than wood.

The saw should be set so the blade extends through the sheet about 1/4 inch. The work must be held firmly down against the table, to avoid chattering and chipping. It should be fed or pushed into the saw slow enough so the cut will not accumulate a ridge of melted plastic about the edge. A bit of practice on the part of the student will indicate the proper rate of feed for a given saw, saw speed, and thickness of stock. Do not try to hurry the cutting by feeding the work too fast, or a poor cutting job is sure to result.

In using a power saw it is possible to cut several sheets of plastic at one time. Of course the saw should be properly set so the blade will protrude through the top of the stack. As in cutting single sheets of plastic, the masking paper is usually left on. Build up of the masking paper adhesive on the saw blade can usually be avoided by occasionally touching a stick of tallow or white soap to the blade lightly, or by feeding a small amount of oil to the blade by means of a wick oiler. If the deposit of the gummy adhesive does build up, the saw should be stopped and cleaned before there is sufficient build-up of the material to cause trouble.

In all sawing, be sure to keep in mind that if the cut edge of the plastic is to be polished, overheating. The coolant may be a 10 per cent solution of any good "soluble" oil coolant. Coolant wets the masking paper and expands it. The wet paper should be removed from the sheet immediately after sawing, if possible. If not, the material should be placed in a drying rack immediately, to allow the paper to dry. If too large a percentage of the soluble oil is added to the coolant, it may react with the adhesive of the paper, so the resulting film will be difficult to remove.

Directing a blast of compressed air into

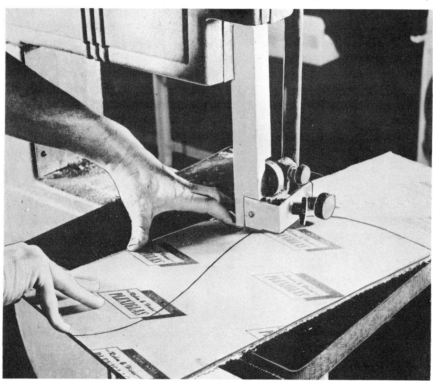

Fig. 7. Bandsawing plastic, using metal-cutting blade.

a good, smooth saw cut will be a great time saver in later polishing operations. Time required to get a saw blade properly sharpened and adjusted is time well spent.

HEAVY DUTY SAWING

Acrylic plastic will not dull a saw blade quickly, but to preserve the life of the blade, it is important to see that the blade does not become overheated. When doing heavy-duty sawing it may be well to spray a small amount of coolant against the blade to prevent

the saw, at the point where the blade contacts the material, will remove the sawdust and help prevent gumming of the blade.

SAWING PLASTIC WITH BAND SAW

A band saw may be used to make either straight or curved cuts in plastic, Fig. 7. Intricate designs and curves of small radii are possible if narrow blades are used. Metal-cutting blades are preferable. To avoid breaking blades, feed stock slowly with only

Cope's Plastics Book

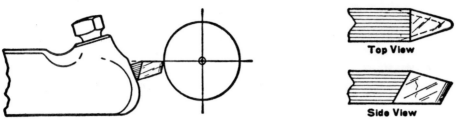

Fig. 8. Lathe tool properly ground for turning plastics. Top of tool is ground flat so there is no side rake, or back rake.

slight pressure. Ease the material back on sharp curves from time to time. The tension of the saw blade should be just enough to prevent the blade from slipping on the wheels. On commercial jobs, a small blast of compressed air directed at the saw blade immediately after it has left the cut, will remove chips of plastic which tend to build up on the wheels of the saw, and help prevent chips from being carried back around the blade.

The guide rolls or blocks should be set so they do not quite touch the teeth yet fully support the rest of the blade. The back-up roll should be adjusted so the blade barely misses it when the saw is running unloaded. The upper guide should be adjusted down to within one-half inch of the work.

USING JIG SAW TO CUT PLASTIC

A jig saw may be used to make an internal cutout without making an entering cut. Such a saw should be used when cutting only one

Fig. 9. Machining plastic in metal-cutting lathe.

sheet at a time. Since the stroke of the jig saw is short, and it does not have a chance to clear itself of chips, it gums up easily. As soon as the saw stops making a clean cut, back off the work and allow the stock and the blade to cool.

A hand-operated coping saw may be used where only a small amount of cutting is needed.

MACHINING ACRYLICS

When we speak of machining plastics, we mean the use of what are commonly called machine tools. These include lathes, milling machines, routers, shapers, drills, threading and tapping machines, etc. We shall devote special sections to drilling and to threading and tapping. In working plastics with machine tools the principle already set forth in sawing holds true. The cutters should not have an actual cutting action--all should be ground so that they have a scraping action with the edge of the cutting tool approximately perpendicular to the surface to be cut. This is true even of drills, as we shall see later in the section on drilling.

Plastics can be turned on a lathe in much the same way and with the sort of tools used in turning brass and copper. See Figures 8 and 9. The cutting tools should have a zero or negative rake, and should be kept sharp and free from nicks or other irregularities. The surface produced by a lathe is semi-matte which can be readily polished by sanding and buffing. As with all cutting operations on plastics, it should be remembered that this thermoplastic material is a poor conductor of heat and tends to soften if much frictional heat is produced in the cutting operation.

In turning, the feed should be light to prevent overheating of the cut and consequent chipping and galling. Roughing cuts 1/16 in. to 1/8 in. deep may be made. The surface speed of the work should be about 500 feet-per-minute. For the final finishing, a cut no more than a 4 or 5 thousandths of an inch deep should be made. (Remember that an average-thickness cigarette paper is about 3 thousandths). This will give a surface that needs no sanding, but can be buffed to a finished polish. The rate of feed should be constant throughout the work to insure an even surface.

Routing may be done on plastics. This is usually done when a special shape is required on the edge of the piece. The cutter must run true and smooth. The cutting edges should have no rake and should operate with more of a scraping than a cutting action.

A milling machine may also be used on acrylic plastic. End mills and horizontal milling cutters do a clean, smooth job of cutting this material, Fig. 10.

Fig. 10. Vertical milling operation being done on Plexiglas.

The surface of a sheet of acrylic may also be planed. This can be done with an ordinary woodworking planer, Fig. 11. The planer bit should be sharpened to a very sharp edge and should be free from nicks. Remember that plastic is somewhat harder than wood, which means that a smaller bite should be taken. The plastic sheet has no "grain" and may be planed from either direction. A power jointer, such as is used in woodworking, may be used also. A jointer should never be used for pieces

shorter than 12 inches. For the shorter lengths of stock, use a hand plane or a file.

If a file is used for removing stock from a piece of plastic it is used in the same way as it would be with soft metal. The ordinary mill-cut, fine tooth file should be used. A well-sharpened hand scraper blade may also be used for smoothing edges on acrylic sheet. Care must be taken not to try to remove too much at once.

DRILLING IN PLASTICS

Using a drill press to drill holes in plastic is shown in Fig. 12.

Twist drills, such as are commonly used for soft metals may be used with plastics. A better finish of the drilled surface and faster work may be done if special drills are used. The chief requirement is a drill that will quickly and efficiently clear the hole of chips.

Bits for drilling acrylic sheet should be carefully ground, Fig. 13. It is especially important that the cutting edge of the drill be ground off to a "zero" rake angle. In other words, the cutting edge should be perpendicular to the face of the material being drilled. The drill should preferably have a "slow" spiral, and the flutes should be highly polished, so that chips will be removed rapidly and efficiently.

A drill that has not been ground off to zero rake will tend to gouge into the work and "grab." This may ruin the piece. In power drilling, the piece should always be held in a suitable clamp, and in such a way that the clamp will not mar the polished surface.

In drilling shallow holes, chip removal is no problem, but in deeper holes accumulated chips will score, burn or tear the surface of the drilled holes if not quickly and thoroughly removed.

Fig. 11. Above. Surface of plastic being planed in thickness planer.
Fig. 12. Below. Drilling in plastic stock is no problem if drill is ground properly. Solution of soapy water makes a fine "coolant."

Working with Acrylics

When drilling shallow holes a few drops of soapy water or liquid soap will do as a drill lubricant.

A watery solution of a "soluble" oil may be used with deeper holes. Large-diameter holes may be more efficiently cut with a hollow end-milling tool. Drills and milling tools may be mounted in a conventional drill press, or

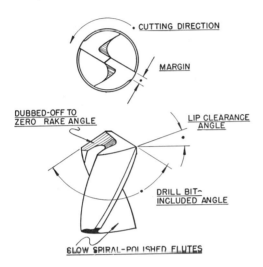

Fig. 13. Drawing which shows how metal drill bit should be ground for drilling in plastics.

they may be used with a flexible shaft, or a portable hand-drill.

The better the equipment used, the better will be the result. Avoid "wobble" of the drilling tool, as any wobble will adversely affect the finish of the hole.

THREADING AND TAPPING

Although it is possible to "thread" (cut threads on the outside of an acrylic rod) or "tap" (cut threads on the inside of an acrylic hole) acrylics, home or school shop work seldom call for this sort of procedure. In most projects it is so easy to cement acrylic plastics together that it is seldom necessary to resort to threading or tapping of the parts to affect assembly. If parts must be frequently assembled or disassembled, the best practice is to use tapped metal inserts.

If threading or tapping acrylics should be desirable, both these operations can be performed with the standard types of tapping and threading tools used for copper and brass, Fig. 14. During tapping or threading, the plastic must be kept cool and use of a lubricant is best in all such operations. The coarsest and heaviest threads possible should be used. No tapped or threaded acrylic part should be subjected to heavy loads or sudden

Fig. 14. Tapping salt shaker in shop of Cope's Plastics Co., with regular machine tap.

stresses, because the threads are likely to give way. The expansion and contraction of acrylic plastic with heat and cold adversely affects the fits of such threads so that they cannot be relied upon for heavy loads or stresses.

SANDING PLASTICS

When the polished surface of a plastics sheet becomes scratched from handling, it may be necessary to restore the surface by finishing it down below the depth of the

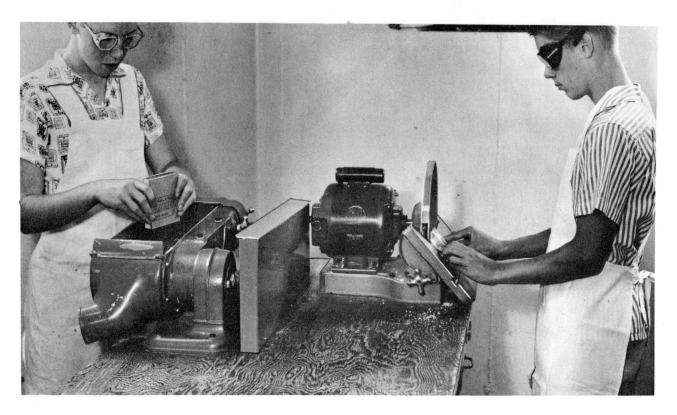

Fig. 15. Disc sanding and wet-belt sanding of acrylic plastic may be accomplished with this setup. The belt is removed if only the disc sander is to be operated. The belt sander is at slight tilt thus allowing for excess water to drain out end spout.

scratch, Fig. 15. If the surface is too deep, it may be impossible to restore the optical quality of the surface except by removing so much material that the surface will be wavy or will have optical distortion. Under such circumstances it may be less noticeable to leave the scratch and use the piece as it is.

Ordinarily, with shallow scratches, or with an edge that has been sawed, milled, or scraped, a mild sanding process, followed by buffing and polishing, will restore the beautiful lustre of the original material.

In any sort of polishing designed to produce an optically clear transparent or reflective surface, the main object is to reduce all irregularities of the surface to the point where they are too small to be seen with the eye. This does not mean that the irregularities are all gone. A surface that appears smooth to the eye will show glaring hills and hollows under high magnification.

When we desire to polish the sawed edge of a plastics sheet, our first move is to change the relatively few and rather deep hills and hollows to a lot more, shallower ones. This is done by first "rough" sanding the material. For this first smoothing operation we use abrasive paper (garnet or aluminum oxide) with a grit rated No. 80-150, if much stock is to be removed, or if the edge or face to be smoothed is quite rough. If hand sanding is to be done, the abrasive paper or cloth should be wrapped around a small block of wood to assure a square edge. Most shops are equipped with motorized sanding discs which may be faced with this coarse grade of abrasive paper, or cloth for the preliminary sanding operations.

An intermediate grade grit should be used next, with grit-size 180 to 220. With a power sander this step may often be skipped, and the sanding job finished by using a wet-sanding belt with grit ranging from 320 to 400. In all sanding operations, remember that you are dealing with a thermoplastic material that gets soft and gummy when hot, whether the heat is generated by friction or by some outside source. Sanding of acrylics works best when the abrasive paper or cloth is wet since this cuts down frictional heat,

Working with Acrylics

and also helps the abrasive grains make a cleaner cut. For the final fine sanding a regular small shop belt sander is the best device for the sanding operation, since it gives a smooth, even stroke of the abrasive against the surface to be cut. This machine should be equipped with a fine-grained wet-or-dry belt. If a one and one-half inch pulley is used on the motor (1750 r.p.m.) and a six-inch pulley on the sanding machine, the ideal running speed is obtained.

After the surface has had its final sanding on the fine-grain belt a smooth, matte surface results. As you progress from coarser to finer sanding, you can actually see the clarity increasing. The hills and valleys in the surface have been so reduced in size that they are barely visible.

If proper precautions have been taken not to press the work-piece too hard against the abrasive paper, and if the whole surface has been uniformly sanded, the piece is now ready for the next step, that of buffing and polishing to bring out the optical clarity of surface which is so desirable in most of our projects with acrylic resin sheets.

BUFFING AND POLISHING PLASTICS

Polishing is usually done in a two-stage operation using motor-driven cloth buffing wheels, Fig. 16. The experimenter must remember that the heat generated by friction, if the work is held too hard against the revolving buffing wheel, will cause the plastic material to overheat. This will result in "burning" or disfiguring the surface that has been so painstakingly brought to this stage of near-perfection.

The cloth buffing wheel is mounted on a shaft, which may be either rigidly mounted in a fixture, or may be on a flexible shaft. The rigid mounted wheel is preferable for small work usually encountered in the school work shop.

The buffing wheel should be of the stitched-cloth type. It has been found that for ordinary school shop use, the best practice is to mount two one-inch, stitched muslin wheels together. This gives a large enough surface to be practical and does not require too much power to turn it. This wheel should run at about 2,000 surface-feet-per-minute -- that is, a 6-inch wheel should turn at about 1,300 revolutions per minute.

A polishing material with a slight abrasive action is first applied to the wheel. This is usually supplied in the form of a stick which is made largely of tallow into which some tripoli has been mixed. Tripoli is a soft, powder-like material of rocky (silica) origin with light abrasive action. Sometimes tripoli is known as powdered rottenstone, diatomite, or kieselguhr.

In this polishing operation the piece of work is held lightly against the revolving wheel. It must be kept constantly moving from side to side while in contact with the wheel. Never hold the piece still because there is danger of "burning" or melting the plastic surface of the material.

The final polish, which brings out the optical clarity of the acrylic surface, is done on a second wheel. It is recommended that this second or buffing wheel consist of two

Fig. 16. Approximate angle at which block of Plexiglas should be held so that buffer makes contact on bottom half of material. Double shaft, 1/2 H.P. motor is utilized, thus permitting use of muslin buffs with tripoli on one side, and flannel buffs with white final polish compound on other side. Buffer is direct drive, 1,750 r.p.m., which is ideal speed for plastic polishing.

conventional one-inch buffing wheels mounted on the shaft together. A buffing wheel consists of stitched layers of cotton flannel, and is softer than a polishing wheel. For most finishing operations a small amount of white rouge is applied to this second wheel. The piece should be moved rapidly back and forth during the polishing operation, and under no circumstances should the work ever be held still while in contact with the buffing wheel.

SOLVENT POLISHING

Sometimes it is desirable to remove the matte finish caused by sanding or (in the case of a drilled hole) by the cutting action

Fig. 17. Grieve-Hendry oven for heating plastics for forming.

of the drill without buffing. This may be accomplished by applying a small amount of acrylic solvent (such as ethylene dichloride) to the matte finish for about 15 seconds--then remove the solvent and allow the surface to dry thoroughly. In solvent polishing, the solvent liquid should never be allowed to come in contact with an already polished surface, as it will pit or otherwise mar that surface. Solvent polishing gives a smooth-looking finish, but it does not have the optical clarity of a sanded and buffed surface.

HEAT FORMING

Acrylic plastic becomes soft and easily bent into any desired shape when it is hot. When the material cools, it stays in the

Fig. 18. Forming of plastic with Hotpack Electric Company's 24 in. portable oven.

shape into which it was formed. There is a small amount of contraction caused by cooling, but except for this contraction, the formed piece is of the same shape given to it when hot. Acrylics are quite rigid and non-formable when at ordinary room temperature, and "cold" forming should not be attempted since this will impose internal strains in the plastic material that will result in "crazing" of the surface, and may cause eventual fracturing of the material.

For hot forming, the acrylic material should be heated uniformly. See Figures 17 and 18. This can best be done in an oven that is thermostatically regulated. On large jobs an oven with forced air circulation is desirable. This latter qualification is not necessary where small parts are being heated, as in home hobby work, or in the school workshop. For these jobs even the home electric range oven, or a thermostatically controlled gas oven will serve very well.

The temperature of the oven should be high enough to heat the material thoroughly so there will be no strain on the formed parts during the forming operation. If temperatures in the oven are too high, the surface of the work may tend to bubble, thus spoiling the optical clarity of the surface.

The best temperature to use will depend on several factors, including the thickness of the material, the method of doing the forming operation, and the shape to which the material is being bent. In considering the heating of acrylic sheets, for example, we must remember, that the material is a poor conductor of heat. It takes a considerable length of time for heat to penetrate entirely through a relatively thick piece. On the other hand, we also must take into account that for the same reason, a thin sheet will tend to cool more quickly than a thick one, and must be hotter in the first place, in order to stay hot long enough for the desired forming operation to be done while the piece is in its thermoplastic state. For example, a sheet of 0.060 in. Plexiglas (type I-A) when heated to 250 deg. F. will cool to below its best forming temperature in less than one minute when exposed to room temperature. A piece 0.250 (one-fourth inch) thick heated to 250 F., will still be slightly flexible after three minutes at room temperature.

If an acrylic sheet is heated too hot, even though it may not be hot enough to cause bubbling, the surface of the sheet may still be so soft that it will suffer small imperfections from the form used (called "mark-off"), or from fingerprints, gloves or specks of dirt that settle on the soft surface.

An acrylic sheet should always be stripped of its masking paper before heating as the heat will cause the adhesive of the paper to "burn onto" the clear surface of the sheet, and will be very difficult to remove.

When placing acrylic sheets in an oven for heating they should rest on a heat-resistant, relatively smooth, flat sheet to prevent sagging or otherwise deforming during the heating process. A sheet of asbestos paper, or glass fiber cloth makes an excellent support for plastic sheets during oven heating. Sheet metal supports are not recommended.

Acrylic plastics that have been "cast" in their original form, have a queer property usually called "plastic memory." If a sheet has been cast flat, and then has been heated, bent, and cooled, it will keep the new shape as long as its temperature stays below about 140 to 180 degrees--depending on the original formula of the acrylic used. If the piece which has been bent or formed is heated above these minimum temperatures for any considerable period of time, it has a strong tendency to return to its original flat shape.

For this reason, formed acrylics have definite limitations on the temperatures to

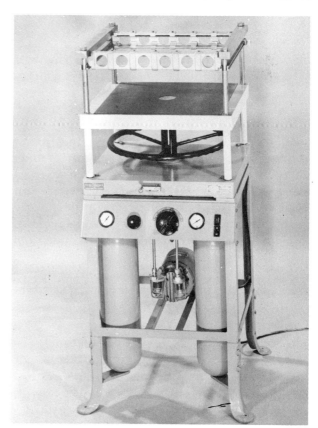

Fig. 19. Di-Acro vacuum-pressure plastic press. This press is used for vacuum forming, stretch forming and pressure forming of thermoplastics.

which they should be exposed. If you happen to make an error in forming a sheet of this material, all that is necessary is to place

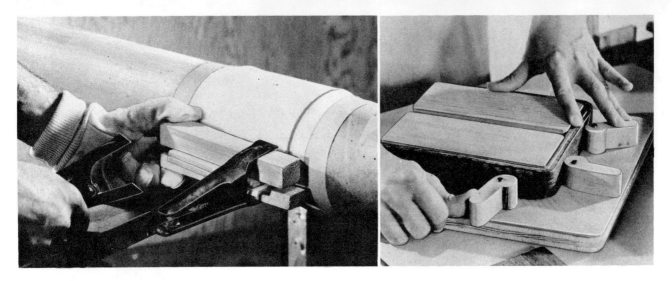

Fig. 20. Left. Cardboard mailing tube makes a practical form for round or cylindrical bends. Strips of wood and spring clamps may be used to hold material against form until it cools. Contact area is covered with a soft cloth such as cotton flannel, to prevent marking soft acrylic surface.

Fig. 21. Right. Wood jig is utilized to make U-shaped bend in plastic. Note off-center pivot blocks used to press against sides thus assuring proper fit and pressure.

the sheet back in the oven, and it will straighten out and be ready for re-forming.

But little effort is needed to form a softened acrylic sheet. A sheet of the hot plastic material is very much like a similar sheet of pure gum rubber in consistency. Many parts can be formed by the use of a vacuum-forming device which withdraws the air from under a sheet of hot Plexiglas placed over a form. The resulting pressure of the atmospheric air on the other side of the sheet forces it into the form. After such a forming operation

Fig. 22. Hotpack Plastiheeter, a strip heater of aluminum construction, used for straight line bends in Plexiglas.

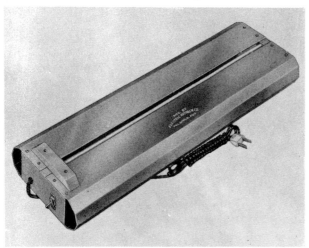

the sheet should be allowed to cool slowly to keep from having parts of the finished product that contain too much internal strain. Note figures 19, 20 and 21.

THE STRIP-HEATER

For simple straightline bends in acrylic sheet, the easiest method is to use a "strip-heater." A strip-heater may be a steam pipe, or a blast of hot air, but usually it is an electrically heated hot strip of some sort. See Fig. 22 and 23. The sheet is supported slightly above the hot strip. The strip should be thermostatically controlled to a temperature between 325 and 350 degrees. The acrylic sheet is placed over the strip-heater, with the heated strip being directly beneath the proposed line of bending. For forming materials thicker than a one-fourth inch (0.250), two heaters should be arranged so that the heating takes place at both surfaces of the sheet.

The acrylic sheet should be kept at least one-sixteenth of an inch from the hot strip for forming sharp bends. To make a bend with a larger radius, a wider strip of the sheet is heated by holding the material farther away from the hot strip. After heating the sheet is bent along the heated line and allowed to cool in the desired shape. It is

Working with Acrylics

usually best to place the bent sheet in a suitable jig to assure that it remains in the desired shape until cool, Fig. 24.

Heat-forming is one of the more delicate operations in treating the acrylics. It requires considerable practice and skill to become highly proficient. The student of plastic-craft should experiment rather broadly before trying to do any of the more complicated projects. An attempt is made, in presenting projects in this book to bring the student up gradually to the more complex projects.

In handling hot Plexiglas sheet from the oven, it is recommended that a pair of soft cotton gloves be worn--both to protect the hands from the heat of the plastic and to protect the plastic from fingerprints. As mentioned earlier in this section, the cooling of Plexiglas sheet--especially the thinner sheets--takes place quickly, so it is necessary to do the forming operation as rapidly as consistent with accuracy and good work.

Simple molds may readily be used in the school shop or home hobby shop, such as a cardboard cylinder for bending sheets on larger radii. Other forms may be readily improvised by the ingenuity of the craftsman. In making more complicated forms for the more advanced worker, or for commercial shops, the manufacturer's literature should be consulted before attempting such work. See Figures 25 and 26.

CEMENTING AND BONDING

Acrylic plastic pieces may be bonded or cemented together in such a way that the joint is almost as clear and transparent as the plastic pieces which are formed together. This work requires skill and patience and it is necessary to have much practice before being able to make the best possible joint. It is not possible, even with the best techniques, to make an invisible joint, because no matter how well the bonding job is done, the changes in the light-bending qualities of the plastic at the joint will show up the joint.

Properly bonded joints will be strong enough for all practical purposes. A well-bonded joint is about three-fourths as strong as the original sheet insofar as "tensile" strength (resistance to being pulled apart) is concerned. This strength factor is true only at room temperatures. At higher temperatures the joint is softer than the original

Fig. 23. Above. Plexiglas being heated with strip heater, prior to bending on straight line.

Fig. 24. Below. After being heated in strip heater, Plexiglas has been placed against wood blocks to form right angle bend. Strip of plywood and weight are used to hold the acrylic in right angle bend until cool.

material and will stretch or tear more easily than the original sheet or piece.

Bonding acrylic surfaces takes place with a somewhat different action than does gluing two pieces of wood together. In the case of the wood-to-wood joint, we depend on the

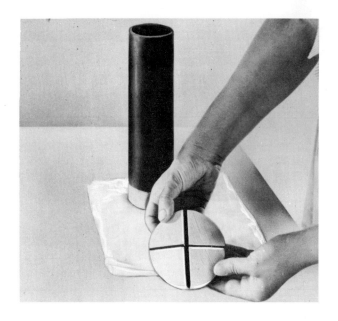
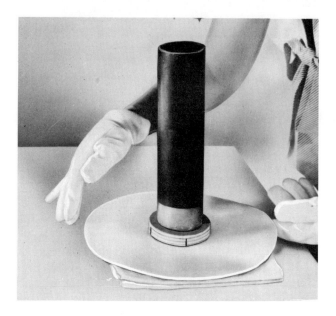
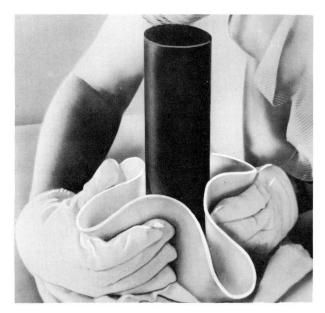
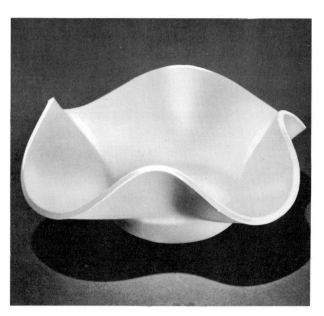

Fig. 25. Simple method of forming an attractive tray from Plexiglas using split block and tube forms.

glue adhering to each of the joined surfaces, thus holding them together. In acrylic bonding, the two surfaces to be joined are treated with a "cement" which has the property of softening the acrylic surface and intermingling with the molecules of it in such a way that when the two surfaces are brought together, the bonding material is actually a polymerized acrylic resin, similar in composition to the original surfaces themselves. Refer to Fig. 27.

The cements used vary somewhat in composition, depending on the kind of joint desired, amount of service strength the joint will need, and the speed necessary in performing the operation. The factor of speed is of considerable importance in commercial shop work.

The manufacturers of Plexiglas list six different formulations of cement, suitable for different purposes. Most of these consist of solvents which will dissolve, (or in cementing work, which will soften) the acrylic material. There are three principal organic solvents known by their chemical names as "methylene dichloride" (M.D.C.), "ethylene dichloride" (E.D.C.), and "vinyl trichloride" (or more properly 1, 1, 2-trichloroethane, and sometimes abbreviated V.T.C.).

For making the strongest possible joint, the surfaces to be bonded are treated with a mixture consisting of one-half methylene dichloride and one-half methyl methacrylate monomer. Remember, the latter material is the monomer (single molecular group) which polymerizes together with other materials to form the original acrylic plastic. When such a mixture of solvent and monomer is used the theory behind the formation of the bond is that the solvent soaks into and, in a sense, opens up "pores" or "voids" between the molecules of the polymerized plastic. Some of the methyl methacrylate monomer follows the solvent into the poly-molecular maze of the body of the plastic--and also the large molecules of the plastic are loosened up enough that when the two pieces are put together the molecules from one piece get mixed up with the molecules of the other piece at the joint. When the solvent evaporates, and especially if this happens in the presence of light and heat, the monomer tends to polymerize, just as it did when the plastic was originally formed. The new joint, is held together with "new" polymerized molecules together with the intermingled huge molecules from the two pieces that got mixed up when the surfaces were softened by the solvent. This, in theory, makes a very good joint--and in practice it has proved to be the best and strongest acrylic-to-acrylic joint.

This cement, containing one-half monomer of methyl methacrylate, takes some time to act--usually from three to fifteen minutes, and hence is slower acting than some of the other cements. Such a joint must be heat-treated to acquire its greatest strength. Heat treatment consists of allowing the joint to stand for eight hours, then to follow that by keeping the joint at temperatures ranging from 122 to 158 degrees F., (depending on the type of Plexiglas or other acrylic polymer being used) for forty-eight hours. This heat treatment drives the solvent out of the joint and also aids in polymerizing the monomer present, thus adding considerably to the strength of the joint, see Fig. 28.

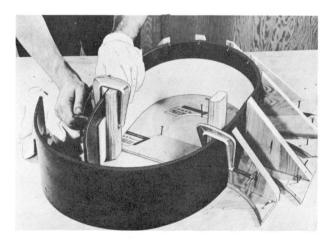

Fig. 26. Strip of heavy leather belting is used to make flexible form for shaping curved portion of project.

Fig. 27. Above. Soaking strip of Plexiglas prior to sealing to another panel. Small nails under edge of strip hold plastic from direct contact with bottom of pan.

Fig. 28. Below. Cut out letters of Plexiglas being cemented to background sheet. Cement is run under letters by capillary action.

Another type of cement sometimes used consists solely of the monomer, methyl methacrylate, without any solvent. This cement is slow acting, but forms a very strong joint when heat-treated. In all the cements using the monomer a very slight amount of a reducing agent, known as "hydroquinone" is added to prevent polymerization of the monomer while it is stored on the shelf. In using these cements, a small amount of oxidizing catalyst must be added just before the cement is to be used. This usually consists of a mixture of benzoyl peroxide and stabilizer--the benzoyl peroxide must be supplied in large enough quantity to overcome the reducing action of the hydroquinone reducing agent, and also to catalyze the polymerization of the methyl methacrylate monomer present.

Most home hobby shop and school workshop projects do not need such thorough and careful bonding. Bonding work such as described in the preceding paragraphs is required when the bonded material must undergo later forming, machining, sanding, etc.

For the less elaborate joining methods such as are recommended for the school shop and the home workshop, solvent bonding is simpler, quicker, and produces a bond of high enough strength for all ordinary uses.

In preparing the joint for cementing, the surfaces if polished, should be left in that condition. If not, they should be sanded or machined until they are smooth and true, and fit accurately. If making a butt-joint of two edges, they should not be first polished because the buffing process usually produces a slight rounding of the corners, thus preventing the square fit needed with this type of joint.

After the joining surfaces are prepared so they fit accurately, one of the two pieces is treated by the "soak" method. In this method the part that is to fit against the other part is soaked by immersion in the cement, or in the solvent (if only solvent is being used). This soaking of the surface to be joined with the other produces a rather deep "cushion" of acrylic material that has been permeated by the solvent or the cement. This "cushion" of softened acrylic plastic should be thick enough so that it goes deeper into the plastic than any small irregularities of the surface to be joined. The time for the soaking operation will vary greatly. Even the manufacturer's recommendations are only a rough guide, and the experimenter should work out an exact soak time for each type of cement and for each kind of cementing problem.

Fig. 29. Left. Sheet of glass being used for soaking. Plastic is placed on pieces of thin wire, cement is injected under edge with an eye dropper. Capillary attraction holds thin, fluid cement in place. After soaking the part is placed in position and allowed to set for several hours.

Fig. 30. Right. Fine wires being used to space two pieces of Plexiglas for injection of solvent. After thorough soaking, wires are removed, and part is allowed to set until sealed.

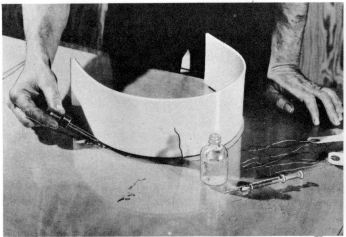

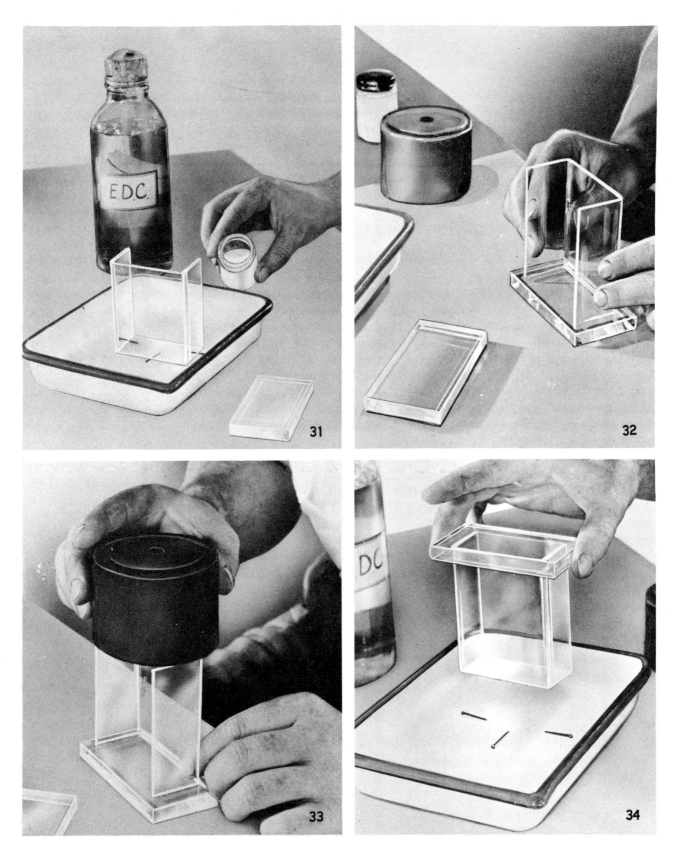

Fig. 31. Formed Plexiglas being soaked for cementing. Fig. 32. Placing soaked edge in position on end member. Fig. 33. Weight is used to hold section in place until dry. Fig. 34. Opposite end is placed in solvent for soaking period.

If it becomes necessary, because of the shapes of the parts to be joined, to restrict the area of the plastic on which the cement or solvent is to act, the part of the surface not to be acted upon should be masked off. There are several methods of thus masking the plastic surface, including a special cellulose or paper tape with an adhesive back, which is applied tightly to the acrylic surface so the tape fits tightly against the surface to be protected, and the cement will not seep under it. A plasticized gelatin solution

Fig. 35. Lining up completed project.

may also be used for painting onto the part of the plastic to be protected from the solvent action of the cement. The masking paper originally on the acrylic sheet is not suitable for masking in the cementing process. A special masking paper or masking solution must be used.

In many cementing jobs where a short, straight edge is to be bonded to another plastic surface, all that is necessary is to support the edge that is to be part of the joint so that it just touches the surface of the solvent or cement.

This may be done by laying glass rods in the bottom of a tray and adding just enough of the cement to cover the rods. A plastic edge laid on the rods will then just be touching the liquid in the tray, and no masking is needed.

Avoid soaking because this makes it take much longer for the joint to harden. If too much of the plastic is softened, it will be squeezed out of the joint when the two parts are brought together, thus giving a joint that will require a lot of cleaning. With a little practice, this method of cementing joints can be done quickly and without splattering cement.

In some joints, it is possible to place fine wires between the surfaces to be joined, then introduce the cement or solvent between the surfaces by means of an oil can, eye dropper or hypodermic needle, Figures 29 and 30. After the solvent has reached all parts of the edges, the wires are removed and the joining surfaces pressed together. In some cases, where the two surfaces to be joined are exceptionally well matched, it is possible to produce an acceptable joint by bringing the surfaces together while dry--then applying the cement to the edges of the joint with eye dropper or hypodermic needle, allowing the cement to spread to all points of the joint by means of capillary action. See photos 31 to 35 inclusive.

In using the soak or dip method of cementing, it is very important to assemble the pieces immediately after removing the soaked or dipped piece from the liquid, before the solvent on the surface has a chance to evaporate. This solvent attacks and softens the dry surface of the other part of the joint, and the excess cement forms a thin cushion on the dry surface of the other piece. The two parts should be held together gently without pressure, for 15 to 30 seconds, to allow a second cushion to form on the formerly dry surface.

Many plastic workers use a cement made by dissolving acrylic sawdust, shavings, or molding powders in ethylene dichloride to make a syrup about the consistency of commerical corn syrup. This cement is excellent for "laminating" two sheets of plastic, and for other applications where fairly large

Working with Acrylics

surfaces are to be cemented together.

Cementing is best accomplished by providing a jig for holding the joint together until it hardens. During the hardening process, the pressure should be heavy enough to squeeze out all air bubbles and to provide complete intermingling of the two softened surface cushions. The pressure should be applied evenly along the entire joint, thus preventing points of stress to develop. The pressure should be kept on to make up for the tendency of the solvent-soaked plastic to shrink during setting or hardening.

USING JIGS

Jigs for holding cemented joints should be constructed so a positive spring, (or other type of) pressure is applied to pull the two cemented surfaces together as the drying and hardening process continues. Pressure should not be great enough to force the cement or cushion out of the joint. This will result in "dry" areas that will not bond. For most joints a pressure of about 10 pounds per square inch has been found to be ample, provided that this pressure does not force either of the parts out of shape. After the joint being cemented has been put in the pressure-loaded jig, it should be inspected to note if there has been any slipping, and if so the pressures should be readjusted. Any excess cement or cushion forced out of the joint should be removed while still wet. This will reduce the clean-up times required in right angle, overlap or rib joints.

Continuous pressure in simple jigs can be provided, for example, by spring clothespins or battery clamps. Under certain conditions the pressure can be applied by compressed air. As an example, if a rib is being cemented into place a jig could be constructed by placing a bar directly over the rib, and applying battery clamps to hold the bar down on the rib every two or three inches. The making of suitable cementing jigs calls for some ingenuity on the part of the experimenter.

Assembled joints should be allowed to stand in their jigs for at least four hours. If the joint must be unusually strong, it should stay in the jig for eight hours. Where a monomer-containing cement has been used, and maximum joint strength is desired, the joint should then be heat-treated for 48 hours, to drive out all the solvent and to hasten the polymerization of the monomer in the cement.

In any event, the joint should be allowed to harden completely before any sanding, machining, or polishing is done. The material forced out of the joint when first brought together tends to shrink until final setting has taken place, and any work done on the joint before that time is sure to leave a "low place" in the surface.

In some instances it may be advisable to thicken the solvent by adding clean chips or shavings of the acrylic material to the solvent to produce a syrup-like cement that can be applied like glue. This sort of cement works on the same principle as the other types, acting as a carrier of the solvent.

HEAT WELDING

It is possible to cause sheets or moldings of acrylic resins to bond together solely by the use of heat and pressure. Such joints are strong and transparent. Heat may be applied to the surfaces to be joined by an electrical resistance strip heated to 660 F. The temperature is critical and should be frequently checked by using a pyrometer. The two edges to be joined are fitted, as in any other type of bonding, then brought at the same time into contact with the heated blade. After about 30 seconds the surfaces in contact with the blade will begin to boil, and possibly to smoke. The heater blade should then be pulled away as quickly as possible and the heated edges forced together at a pressure of about 150 pounds per square inch. Pressure must be maintained until the joint cools below the softening temperature. Also, the joint should be cooled as slowly as possible to keep from setting up undue internal stresses in the material.

FRICTION WELDING

The heat of friction may also be used to heat-weld acrylics together. One simple

51

application of this method is that of joining the end of a round rod to a flat sheet. In this operation the end of the round rod is machined off flat. It is then fastened into the chuck of a lathe or drill press so that it can be rotated at about 1000 r.p.m. The flat end of the rod is forced squarely against the surface of the sheet, which must be held firmly in place. The rod, in contact with the sheet, is rotated and pressure is applied. Enough heat caused by friction is generated to depolymerize the rubbing molecules and thus fuse the parts together at the point of contact. Rotation should be stopped instantly, when the proper temperature is reached. The pressure is increased, and the pieces are held firmly in place until the joint cools. Clear, bubble-free joints may be made by this procedure in a few seconds. Friction welding of flat surfaces, or of other areas than circular sections may be done, but special fixtures must be devised for this type of work and it is not recommended for the school shop or home hobbyist.

CLEANING—ANTISTATIC TREATMENT—WAXING

After assembling a plastic project, the surface of the material should be thoroughly cleaned in order to give the finished piece the best possible optical clarity. All fingerprints, grease, remnants of masking paper adhesive, etc. should be removed from the surface to provide the clear, sparkling clearness of the original surface.

In most instances, the plastic can be effectively cleaned by washing with lukewarm water and a pure soap or synthetic detergent. Attempting to clean plastic surfaces by dry rubbing is likely to result in scratching the surface, even slightly, with accumulated gritty dust or other accumulations on the surface. If cleaning without water is necessary, dust carefully with a damp soft cloth or chamois, wiping the surfaces gently.

DO NOT use dirty cloths, boiling water, or any strong solvents such as alcohol (except isopropyl), acetone, carbon tetrachloride, etc., or any window-cleaning fluids which may contain such solvents. Several of the common window-cleaning fluids on the market contain carbon tetrachloride in emulsion form. This is an active solvent for acrylic resins.

After washing the plastic, remove excess water by blotting the surface, using a slightly damp cloth or chamois.

If there is much grease or many fingerprints on the plastic surface, or if some adhesive from the masking paper has been left on, it may be necessary to use a grease solvent. Such cleaners must not have any solvent reaction toward the acrylic material. Cleaners coming in this category include the hydrocarbon solvents which evaporate quickly and leave no residue upon evaporation. Satisfying most of these demands is mineral spirits, a relatively slow evaporating liquid with a high flash point. (The flash point is the temperature at which vapors coming off the material, as it is gradually heated, will give a flash when a lighted taper is passed across the surface of the liquid). There is also VM&P (varnish-makers and painters) naphtha, which evaporates quickly, but has a low flash point. (The flash point of any flammable liquid is a direct measure of the fire hazard of the liquid--that is, the lower the flash point the more likely the liquid is to catch fire when in use). These hydrocarbon solvents should be used by slightly moistening a soft cloth with the solvent (solvent should be dispensed from a safety can, to avoid danger of fire). VM&P naphtha makes an excellent cleaning fluid for cleaning acrylic surfaces before painting. After using a hydrocarbon solvent, the surface, except that which is to be painted, should be washed with soap (or detergent) and water.

For ordinary small shop work and for the home hobbyist, the authors recommend cleaning with a soft cloth moistened with common isopropyl alcohol (not over 70% alcohol) "rubbing compound" obtainable at any drug store. This material is not likely to cause later "crazing" of the edges of the acrylic material such as will occur if 90% isopropyl is used. The 90% isopropyl is recommended by the manufacturers of plastic for cleaning surfaces, with the provision that the surface so cleaned, be later heat-treated to prevent danger of crazing of the cleaned surface or the edges of the surface. Our experience has been that the 70% isopropyl alcohol compound, although not quite as good a grease or fingerprint solvent as the more concentrated form, will not cause this crazing of the plastic surface, if used and wiped off immediately.

ANTISTATIC TREATMENT

Most plastics are decidely nonconductors of electricity. A nonconductor of electricity is called an "insulator" or "dielectric." Such a substance is defined as one on which an electric charge is not communicated from one part to another. Friction tends to build up charges of electricity (static electricity). The more nonconductive the substance is, the more the tendency for this charge of static electricity to build up, simply because it is not conducted away. Consequently most of the plastics, and for our purpose, particularly the acrylics, tend to build up large charges of static electricity on the surface. The peeling off of the masking paper builds up a lot of static. Notice, that as the paper is peeled off, dust particles, paper fragments, etc., seem to fly to the clean surface.

Washing with water reduces this static temporarily. Washing or cleaning with 70% isopropyl alcohol, also tends to remove the static, but it quickly reappears as soon as the alcohol dries. Some substances tend to form a permanently antistatic layer on the surface of acrylics. Most of the waxes when applied in a very thin coat help to reduce static accumulation. A compound called Wilco Plastic Cleaner will impart antistatic properties. There are other similar compounds on the market. Almost any good commerical grade wax will impart a certain amount of antistatic properties, and in addition will give the finished part a high gloss by filling in minor scratch marks. It will also help prevent mild surface scratching of the optical surface. Wax should be applied in a very thin coat with a soft, clean cloth, and polished lightly with clean cotton flannel or jersey. After polishing with wax, going over the piece lightly with a damp cloth will ground any remaining static charge and prevent dust accumulation on the finished surface.

Examples of depth carving in solid Plexiglas blocks, by Brown Plastics Co. Because of the many reflections in the surfaces of the crystal clear plastic, a rose can become a "bouquet" if set at an angle.

PROJECTS FROM PLASTICS

The projects described in this section deal chiefly with acrylic materials. A few projects using other types of plastics are included in this book, some of them in this section on acrylics. Differences in procedures required to handle the other plastic materials is included with the descriptions of the projects.

One important point that must be mentioned here is that when a project calls for bonding acrylic resin to other material, a good universal bond cement should be used. For example, if findings such as pin backs are to be fastened to plastic, do not try to use the usual solvent-type cement that is used in bonding two pieces of plastic together. Rather, use one of the types of cements such as are put out for cementing model airplanes together, or any other good household or all-purpose type of cement. It is necessary that the bonding material used have a certain amount of "give" since the rate of expansion with changes in temperature is different for different materials.

Each project consists of an illustration of the completed project, together with a written description. There will also be a list of the materials needed to make the pictured article, and a step-by-step description of procedure. In some instances there are line drawings of parts to show relative sizes, outlines of curvature, etc.

Dimensions are given for the various parts of the projects, but many of these may be varied at the discretion of an experienced craftsman. However, if you are working in a school class, be sure to consult your instructor before changing any dimensions.

You will notice that the dimensions given on many of the projects appear to be odd-- 11-7/8 inches for example, instead of 12 inches. This is done to get the most economical lengths or sizes of plastics from the sheets of stock. Plastic sheets generally come in widths of even foot multiples. If a one-inch cut is made across the end of a 36 inch sheet, it would be impossible to cut three 12 inch strips from this since the saw cut would remove about one-eighth inch at each cut. If we aim at cutting strips only 11-7/8 inches long, we have made ample allowance for wastage due to the saw cut, and for any further removal of material that may be required for sanding and polishing.

When cutting odd-shaped pieces from acrylic stock, it is recommended that the student make paper patterns of the various pieces, and fit these pieces onto the plastic stock to avoid excess waste. Also, these patterns will enable the student to mark the outlines of the various pieces on the masking paper of the plastic stock before cutting.

No attempt is made in this book to teach the student the basic facts of tool handling and maintenance. It is understood that these facts have been taught in previous courses, or that the necessary information will be provided by the instructor.

Included in a number of the photos are projects other than the one covered by the step-by-step procedure. These alternate designs have been included as an incentive to the student to create his own designs in plastics.

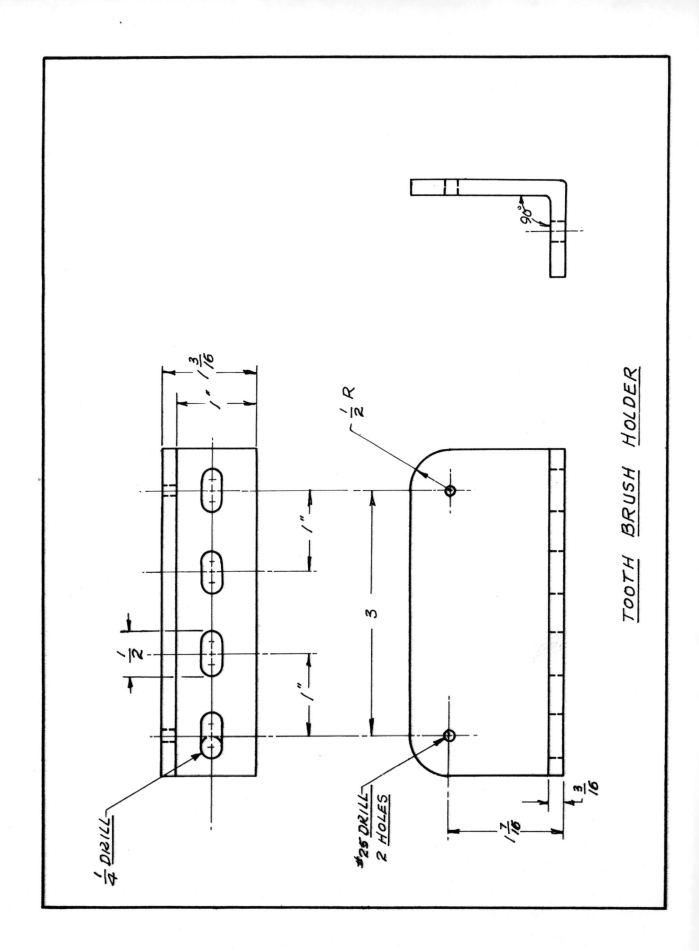

TOOTH BRUSH HOLDER

SUBMITTED BY:
Edison Clark
Sunnyside Jr. High
Lafayette, Indiana

In starting to work with plastics, it is advisable to begin with a simple project like this Tooth Brush Holder.

The project is well suited for elementary students.

MATERIAL REQUIRED:
1 piece clear plastic 3/16 x 2-3/4 x 4 in.

PROCEDURE:
1. Cut the plastic to size indicated on drawing.
2. Lay out corners and holes.
3. Drill holes using bits of proper size.
4. Use coping saw and cut out center section between pairs of holes to form slot. File smooth.
5. Sand and polish all outer edges, and solvent polish hole edges.
6. Remove all masking paper and clean plastic.
7. If strip-heater is available, heat on line indicated and fold at 90 degrees. Hold in position until cool. If oven is used, heat for approximately five minutes at 280 F. then form at right angle at indicated line using two blocks of smooth surfaced wood covered with soft flannel. Hold in blocks until cool.
8. Clean and wax.

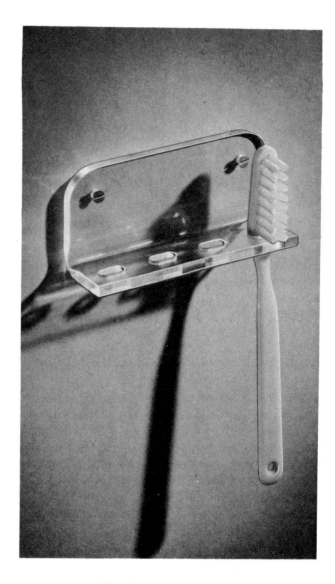

Tooth brush holder.

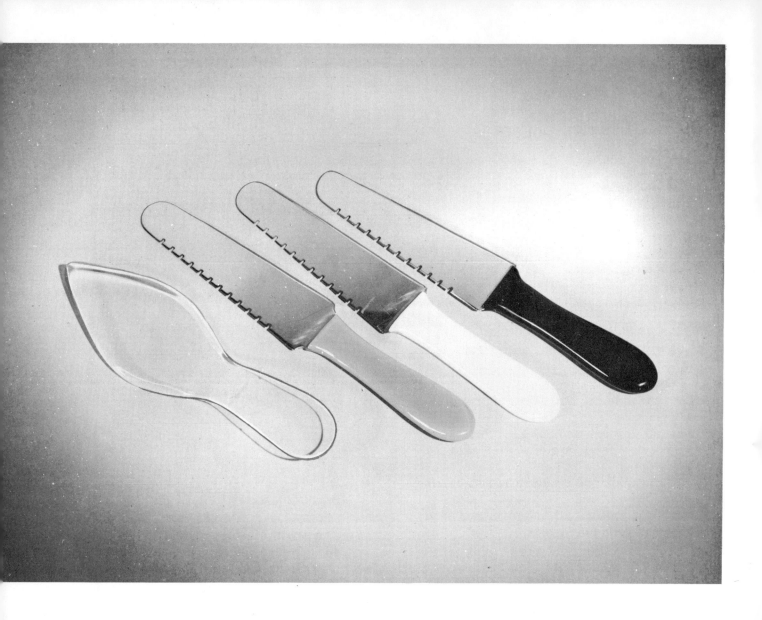

PLASTIC PIE. CAKE SERVER

SUBMITTED BY:
  Lloyd Schmidt, Allison School
  Wichita, Kansas, and
  Edison Clark, Sunnyside Jr. High
  Lafayette, Indiana

The Pie and Cake Server, shown in the photo at the left makes an excellent elementary project. The other servers shown in the photo (laminated) are intended as alternate design suggestions.

MATERIAL REQUIRED:
1 piece clear plastic 1/8 x 2-3/4 x 8-1/4 in.

PROCEDURE:
1. Trace pattern onto plastic.
2. Saw out, using either hand, or power saw.
3. Bevel edges of blade on top side, back from point to beginning of handle. This may be done with a file, scraper, or by sanding on belt wet-sander.
4. Wet-sand and polish the edges of the entire piece, including bevel.
5. Heat flat piece in oven about 280 degrees F. until plastic is soft enough to bend.
6. Bend up handle on server, by hand, as shown in side-view of drawing. Hold in place until cool. Clean and wax.

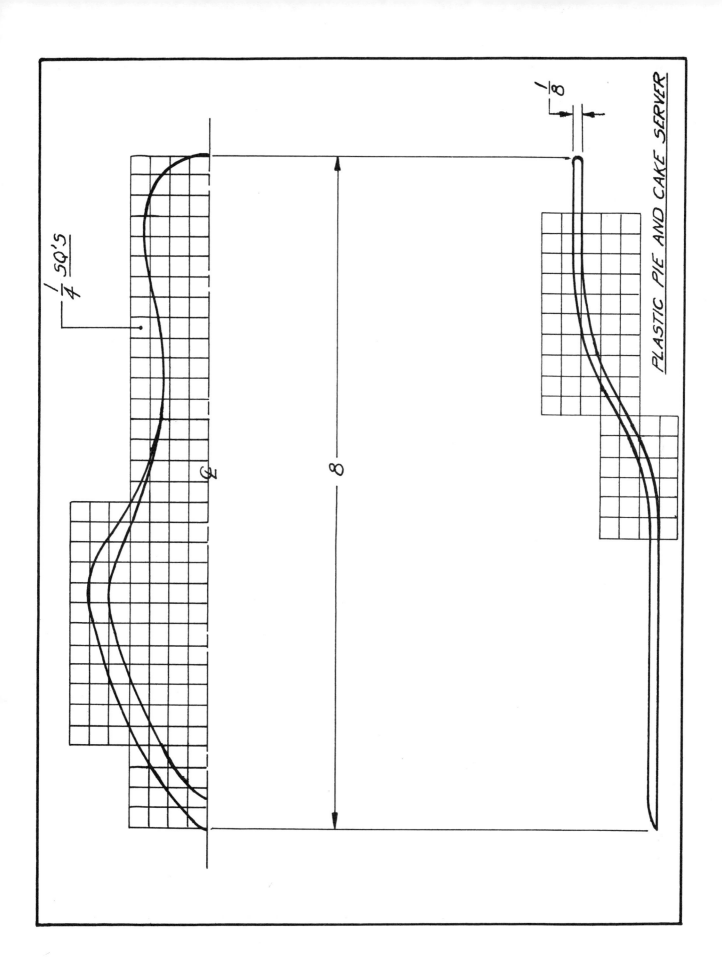

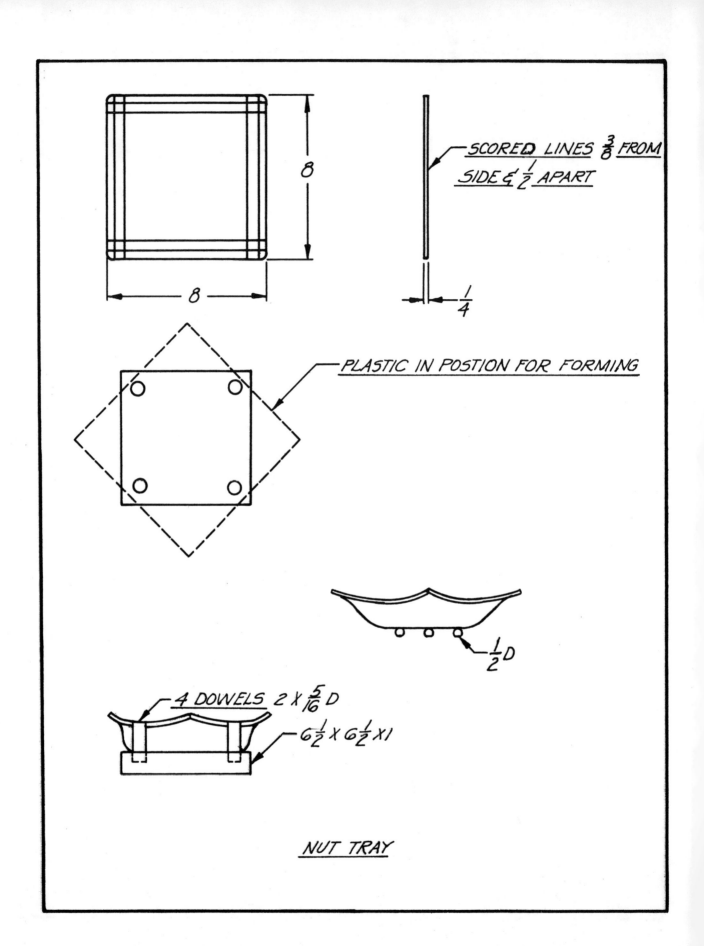

NUT TRAY

SUBMITTED BY:
D. Shaddy
Jennings Jr. High School
Jennings, Missouri

This simple tray is decorative as well as useful.

The polished edges of the curved plastics sheet illustrate the light-piping properties of the acrylics. The entire piece glows and sparkles with the lighting effects produced.

MATERIAL REQUIRED:
1 square sheet clear plastic 1/4 x 8 x 8 in.
3 polished plastic balls 1/2 in. diameter
1 forming jig (optional)

PROCEDURE:
1. Remove masking paper from one side of plastic sheet.
2. Score decorative, parallel lines on unmasked surface. This may be done by hand, using a ruler and a sharp-pointed instrument such as a scriber, or a sharp ice pick. (This will be the <u>under</u> side of the tray).
3. Remove remainder of masking paper. Clean surfaces of all adhesive particles. If necessary, rub surfaces with soft cloth slightly moistened with 70% isopropyl alcohol. Wipe dry immediately. Sand and polish edges of sheet. Do not round off edges. (Rounded edges do not have sparkle and light-emitting qualities of square edges). Slightly round off sharp corners, only. The sheet is now ready for heating.
4. Make form for shaping plastic sheet. Cut block 5 in. square from 2 in. wood plank. Round corners of block to small, even radius on all four corners.
5. Heat plastic sheet in electric (or other controlled-heat) oven at 275-280 degrees F. for three to five minutes, or until sheet becomes pliable.
6. Drape sheet over forming block placed on clean table. Handle hot plastic with soft cotton gloves. Bend sheet at middle of edges as shown in photo, using thumbs and fingers of both hands. Hold formed sheet in desired shape until cool (about 3 minutes). NOTE: For more accurate forming, prepare jig by driving wood dowel pins into holes in a board. For size tray shown here, set dowel pegs in square so that diagonally opposite pegs measure 6-3/4 in. from inside of peg. Place forming block (Step 4) on top side of hot plastic sheet. Force sheet and block down over dowel pins. Small weight on forming block holds piece in place until cool.
7. After piece has cooled to room temperature, attach feet. These are made by sanding a "flat" on each of three plastic balls. Apply cement to flat spot and bond to bottom of tray. (Three feet will stand flat on any surface--four may not).
8. Clean finished piece with soft cloth slightly moistened with 70% isopropyl alcohol. Wipe dry immediately. Wax all surfaces of piece with thin coat of good paste or other wax. Polish with soft cloth. (NOTE: Wax tends to fill in minor scratches on polished plastic surface, making finished article appear clearer and shinier than if left unwaxed).

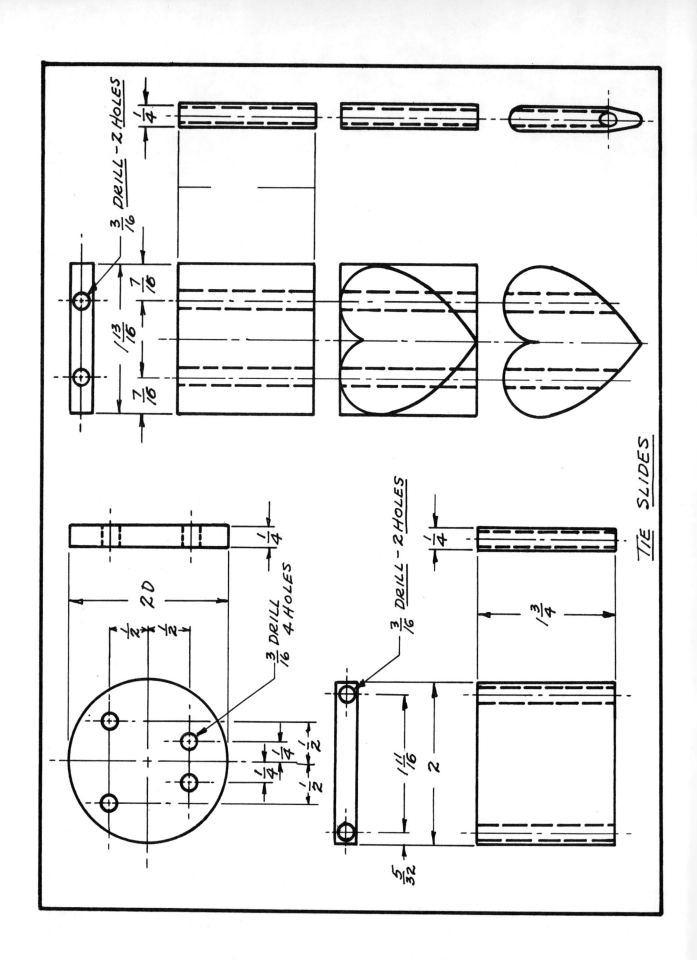

TIE SLIDES

SUBMITTED BY:
Lloyd Schmidt
Allison School
Wichita, Kansas

The simple tie slides pictured here are easily made from sheet plastic. In the photo are shown three patterns. The lower slide has an internally carved rose figure. The center one is a red heart-shaped design. The slide shown at the top of the photo is a circular-shaped piece of clear plastic with a cameo cemented to the face. In two of the designs, the holes for leather thong of the slide are drilled through the length of the blocks. In the upper one, the holes are drilled perpendicular to the face of the plastic block and the thong is threaded through. The following directions are for making the round slide.

MATERIALS REQUIRED:
1 square of clear plastic 2 in. x 2 in. x 1/4 in. thick
1 cameo decoration
1 leather thong of suitable length

PROCEDURE:
1. Draw the outline of the desired circle on the masking paper of the plastic square.
2. Saw out circle, using hand coping saw with fine-toothed blade, or powered jig saw.
3. File or scrape sawed edges until the circle is perfectly round.
4. Mark locations of four holes, then drill holes using a drill of size so leather thong will fit snugly when pulled through hole.
5. Remove masking paper from both faces of plastic sheet. Clean with cloth slightly moistened with 70% isopropyl alcohol and wipe dry immediately using soft cloth.
6. Wet-sand and polish sawed edge of circle.
7. Solvent-polish holes. This is done by moistening roughened inner surface of hole with acrylic solvent, such as ethylene dichloride. After 15 seconds wipe the inner surface of hole dry with soft cloth.
8. Cement cameo decoration onto face of clear plastic circle. Allow sufficient time for cemented joint to set thoroughly. (Suggested time 3 to 4 hours). Then wax and add final polish with soft cloth.
9. Thread leather thong through holes as shown in photograph bringing both ends of thong up through top holes from back side, then down through both lower holes from face side.

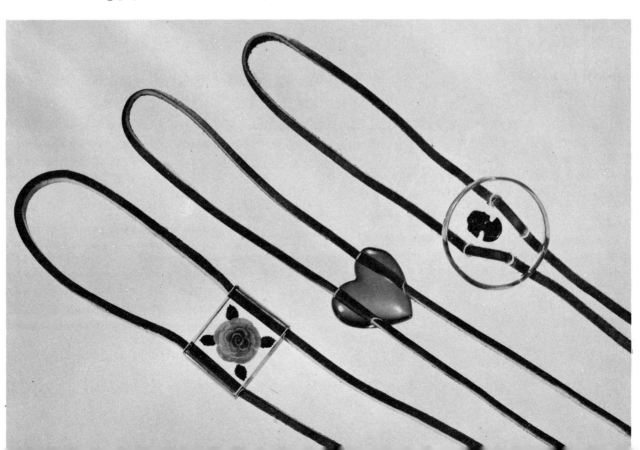

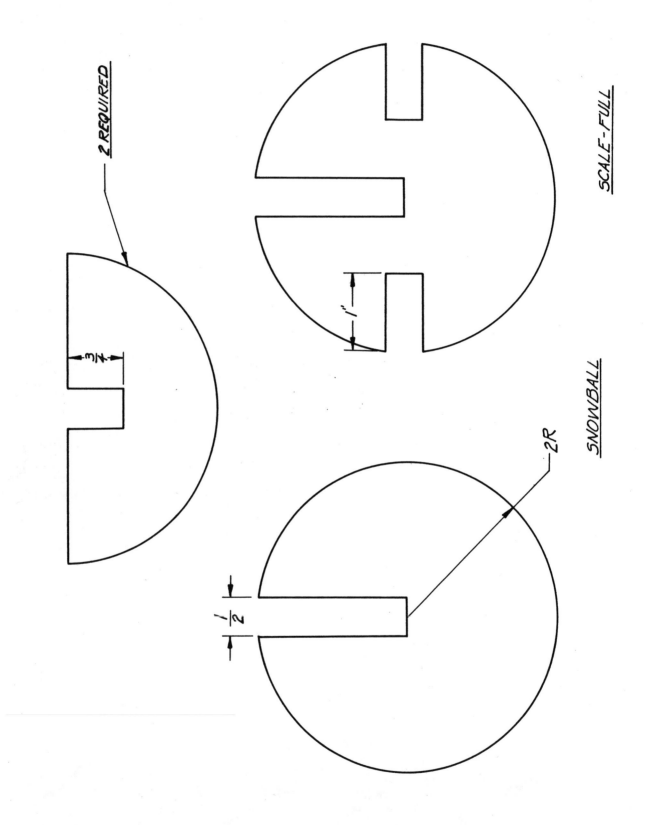

STYROFOAM ORNAMENTS

SUBMITTED BY:
Lloyd Schmidt
Allison School
Wichita, Kansas

The projects shown here are made of expanded polystyrene, "Styrofoam." No attempt is made to furnish patterns for all of the pictured items, but the technique of fitting the three-piece snowball at the top will be worked out, and full-size patterns provided. Since Styrofoam is so easily cut with knife or hand saw, it will be no problem to easily cut it to shape. Parts should be cut and fit before cementing, then should be cemented with a proper type of cement which will not ruin the Styrofoam, such as Rez-N-Glue.

MATERIAL REQUIRED:
(for one Styrofoam snowball)
1 piece 4 x 12 x 1/2 in. white Styrofoam
1 pipe cleaner for hanging

PROCEDURE:
1. Trace pattern of each of the parts onto the Styrofoam.
   NOTE: Diagram "B" requires two pieces.
2. Using hand coping saw, power band or scroll saw, cut the units to shape. Cut should be made accurately on the 1/2 in. wide slots as these should fit snugly over the 1/2 in. thick material. Should the material be thicker or thinner than 1/2 in. cut the slots accordingly.
3. If outer edges of circular cuts are irregular, this may be sanded evenly on disc sander set at 90 degrees.
4. Fit all four units together, "A" and "C" first, then follow with the two units of "B." These will fit together and stay without gluing if properly fit.
5. Insert pipe cleaner into one end, thus making a hanging snowball ornament.

NOTE: Patterns may be made for other items such as the star design or the Christmas tree design by folding a piece of wrapping paper, and cutting a half design, then unfolding for full pattern. To make an item such as the Snowman pictured, you will need white Styrofoam balls, a pipe cleaner, and scrap pieces of Styrofoam. Colored Styrofoam can be utilized in the snowball described above or in any of the ornaments, thus lending itself well to Holiday colors. White Styrofoam may be painted, using any water base paint.

Ornaments made from Styrofoam.

Projects in this book have lots of money-making possibilities.

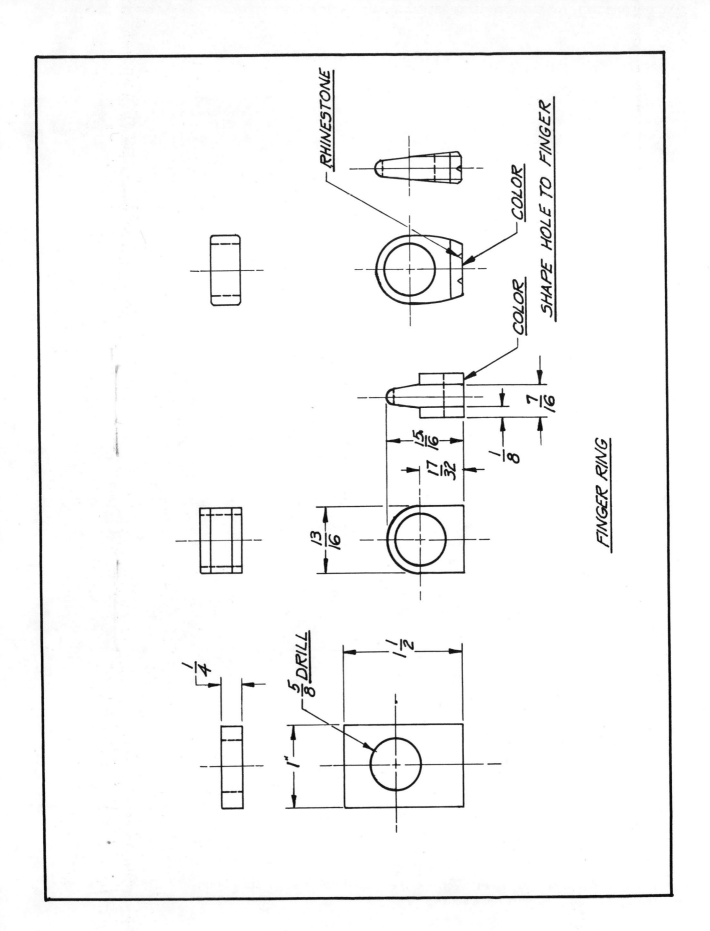

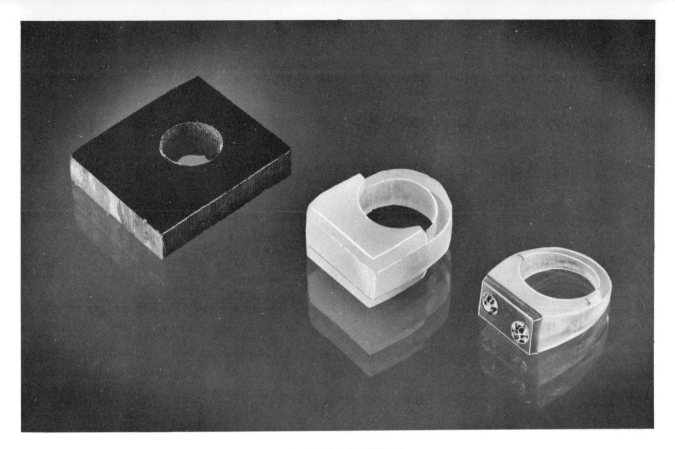

PLASTIC FINGER RINGS

SUBMITTED BY:
Lyle Carter
Northeast Missouri State College
Kirksville, Missouri

Making of plastic rings offers all kinds of opportunities for students to turn out small colorful projects.

The photo shows the three steps in laminating the ring, and shows that a variety of colors can be used.

MATERIAL REQUIRED:

Almost any small scrap pieces of plastic generally available in the shop. Base center piece of ring is generally from 3/16 in. to 1/4 in. thick and can be clear or colored. Side laminations can be of a contrasting color which will give a pleasing array. Crystal rhinestones from inexpensive costume jewelry may be used to add to the attractiveness of the project.

PROCEDURE:
1. Secure piece of clear or colored plastic about 1 x 1-1/4 x 1/4 in.
2. Cut contrasting color of plastic in size about 1 x 1 x 1/8 in. (2 pieces required-- one for each side).
3. Cut third contrasting color about 5/8 x 1 x 1/8 in. thick.
4. Remove all masking paper from plastic surface.
5. Using soak method, seal pieces together as shown in photo. Dry thoroughly.
6. Lay out location of cross drilled hole with scriber.
7. Drill hole approximately 5/8 in. diameter through block. This hole may be enlarged to size desired to fit wearer.
8. Sand and smooth all edges to shape as desired. Round all sharp inner edges.
9. Using regular drill bits, drill two holes evenly spaced in top section to receive rhinestones. Size of holes will depend on size of rhinestones.
10. Using "Rhinestone Cement" (a special cement which does not affect the foil backing of rhinestones) glue stones into place and allow to thoroughly dry.
11. Polish all surfaces to high gloss.
12. Clean and wax.

ARM BRACELETS

SUBMITTED BY:
George Feuerstein
Winsted, Connecticut, and
Lloyd Schmidt
Allison School, Wichita, Kansas

Plastic bracelets may be made in a great variety of shapes, widths, and sizes.

Five design suggestions are shown in the photo. Step-by-step procedure for making the bracelet shown at the right of the photo--a typical bracelet making job follows.

MATERIAL REQUIRED:
1 piece of black plastic 1/16 x 5-1/2 x 3/4 in.
1 piece clear plastic 1/4 x 3/4 x 3/4 in.

PROCEDURE:
1. Cut plastic to size.
2. Shape ends of bracelet to desired curvature (free form).
3. Bevel the 3/4 x 3/4 in. piece at 30 degrees to full feather edge.
4. Sand and polish both pieces.
5. Remove maskingpaper from all surfaces.
6. Internally carve the 3/4 x 3/4 in. piece with suitable design. (Red roses will go well with black band).
7. Heat the strip of black plastic in oven for about three minutes at 275 degrees F. and form around block of wood cut to wrist shape, or, if desired you may place cloth around wrist, and form bracelet to wrist shape. Form for rather loose fit.
8. Using soak method, seal the carved square to the center of wrist band and allow to dry thoroughly.
9. Clean and wax.

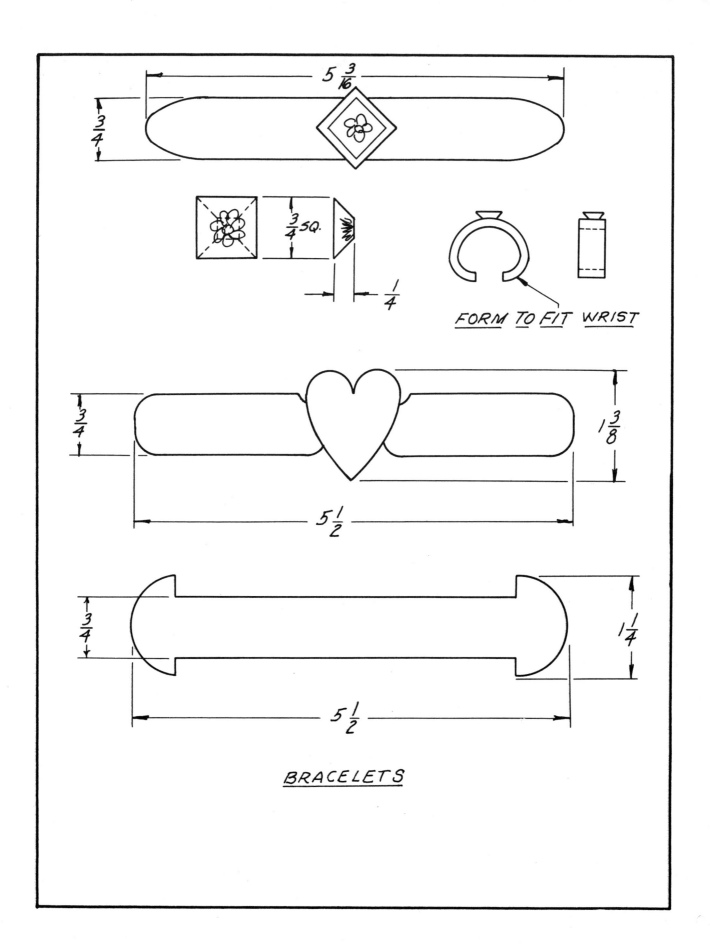

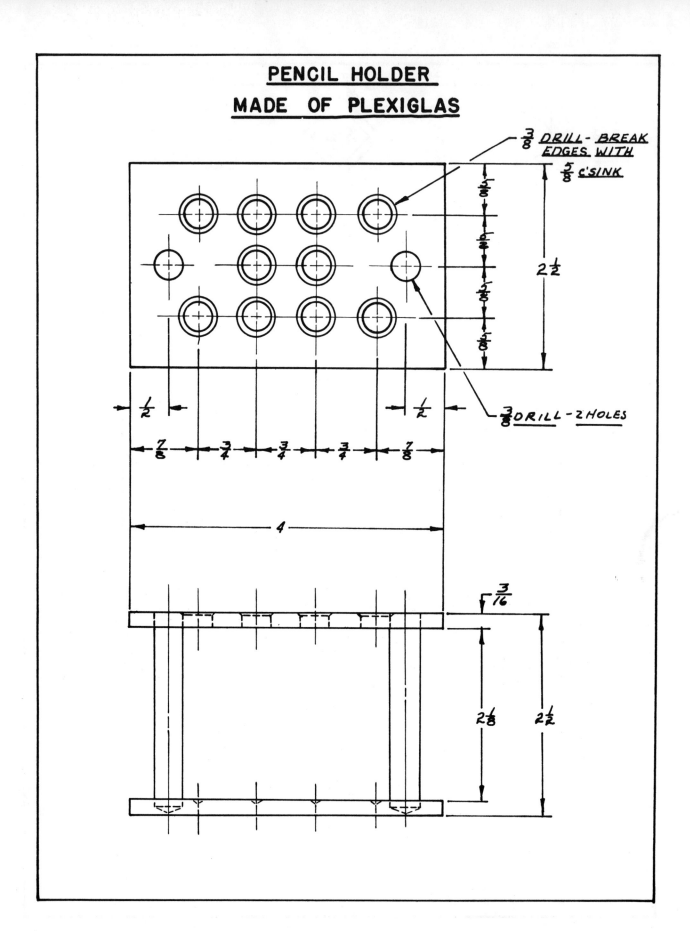

PENCIL HOLDER

SUBMITTED BY:
Edison Clark, Instructor
Sunnyside Jr. High School
Lafayette, Indiana

A simple but useful project.

MATERIAL REQUIRED:
2 pieces 3/16 x 2-1/2 x 4 in. clear Plexiglas
2 pieces 3/8 dia. x 2-1/8 in. long clear cast Lucite rod

PROCEDURE:
1. Cut the two pieces of clear Plexiglas sheet to size, 3/16 x 2-1/2 x 4 in. Also cut 2 pieces of clear cast Lucite rod to length indicated.
2. Wet sand and polish edges of Plexiglas pieces.
3. Square off ends of clear rod carefully against disc sander.
4. Tape the two pieces of Plexiglas together, lay out drilling points, and drill 3/8 diameter holes through top plate at points indicated, and just slightly into second piece of material. This "spot drills" the second piece, so that when assembled it forms a resting point for pencils.
5. Lightly countersink the 3/8 in. diameter holes with a 5/8 in. drill on top side of top plate only.
6. Solvent-weld the two posts into position between the plates as indicated.
7. Clean and wax.

STYROFOAM CUTOUTS

Working with Styrofoam has lots of interesting possibilities. Any number of Holiday decorations can be made from this material. Styrofoam may be cut with coping saw, scroll saw (by hand or powered) or with a sharp pocket knife. Two pieces of the material may be rubbed together to smooth off edges, thus the material becomes its own "sandpaper." Pieces of Styrofoam may be glued together with Rez-N-Glue cement.

Patterns are provided for a Cupid and Heart Valentine cutout, and an Easter Rabbit and Egg decoration. Since the Styrofoam may be purchased in white, green and red, an attractive combination of color can be made.

MATERIAL REQUIRED:
(Sufficient to make the items pictured)
1 sq. ft. of 1/2 in. thick red Styrofoam
1 sq. ft. of 1/2 in. thick white Styrofoam
1 sq. ft. of 1/2 in. thick green Styrofoam
1 pc. scrap Styrofoam 2 in. x 2 in. x 2 in. (for shaped egg)
6 pipe cleaners
4 pcs. wood dowel 3/4 in. dia. x 3/8 in. thick

PROCEDURE:
1. Make paper patterns of designs to be used.
2. Trace patterns on Styrofoam, using soft lead pencil.
3. Cut Styrofoam to shape with saw or knife.
4. Finish edges by rubbing a piece of scrap Styrofoam against edge of project.
5. Using Rez-N-Glue, seal patterns to bases as desired.
6. Insert pipe cleaners into proper positions. Cleaners may be used for axle of egg cart, and glued into holes in wood dowels which serve as wheels.
7. Eyes of rabbit may be short pieces of pipe cleaner or sequins held in place with short pins. Rabbit whiskers and mouth are made of the cleaners also, glued in position.

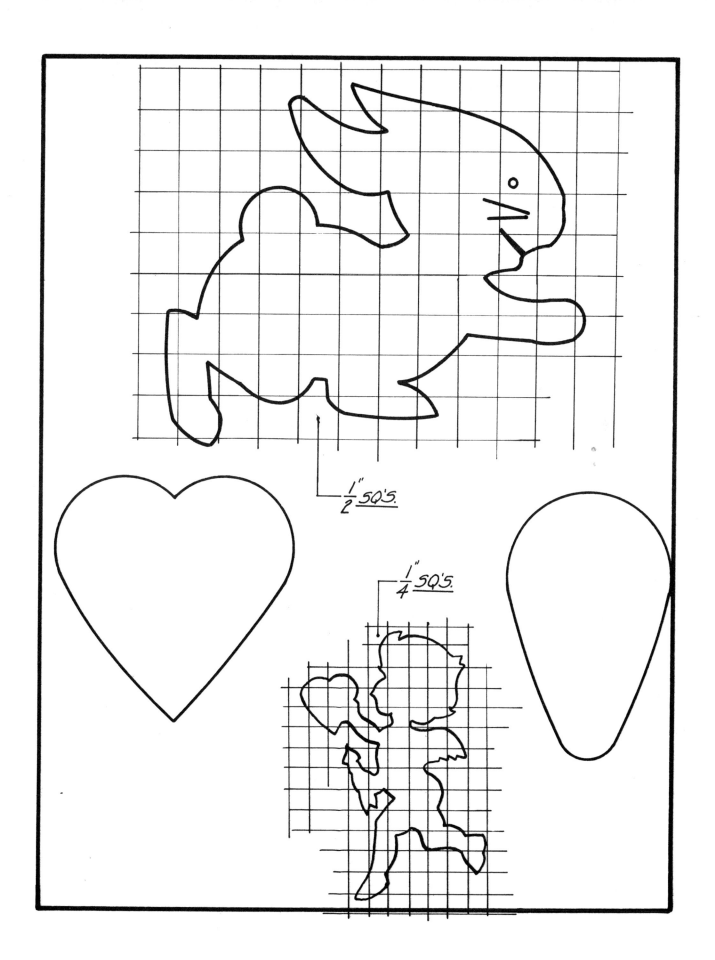

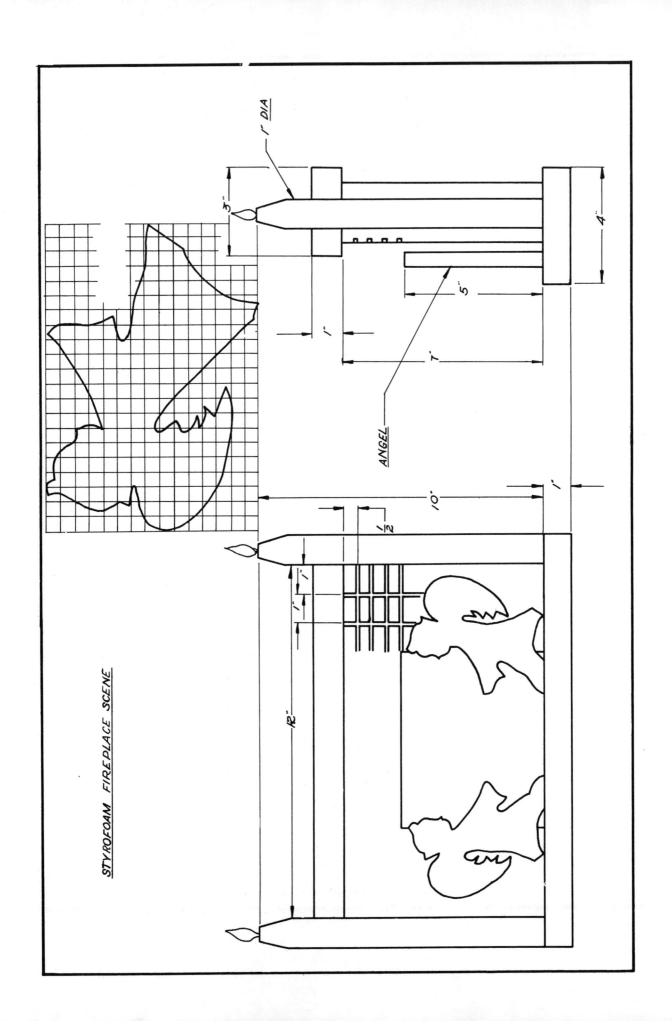

STYROFOAM FIREPLACE SCENE

SUBMITTED BY:
Lloyd Schmidt
Allison School
Wichita, Kansas

This is truly a unique project. It is constructed throughout of Styrofoam except two red pipe cleaners which represent candle flames. Cost of making this project is just a few cents, and yet it presents a very striking ornament for the Christmas Holiday Season. It takes only a short while to complete the project.

MATERIAL REQUIRED:
1 pc. white Styrofoam 2 x 7 x 12 in.
1 pc. white Styrofoam 1 x 3 x 12 in.
1 pc. white Styrofoam 1 x 4 x 12 in.
2 pcs. white Styrofoam 1 x 1 x 10 in.
2 red pipe stem cleaners
A small amount of red Styrofoam paint
A small amount of Rez-N-Glue cement
2 pcs. white Styrofoam 1/2 x 4 x 5 in.

PROCEDURE:
1. Cut the block of 2 in. Styrofoam to size, 7 in. x 12 in. Using a soft leaded pencil, draw in the 5 in. x 6 in. center cut. Cut out this center on band saw.
2. Cut top plate and bottom plates to size (1 in. x 3 in. x 12 in. for top and 1 in. x 4 in. x 12 in. for bottom). These are white Styrofoam.
3. Using 1 in. square bars of the white Styrofoam, shape them to a round bar 1 in. in diameter, by using a block of scrap Styrofoam as a sanding block. Thus, by rubbing this block against the corners and edges of the square bar, it will cut away the material to the desired shape. Shape and taper per drawings.
4. Insert pipe cleaner (candle fire) in upper end of each candle.
5. Using any water based paint, or regular Styrofoam Paint, which comes in pressurized cans, paint the inside cutout of the 2 in. main block, and also spray one front surface, and the two end surfaces. Allow to dry.
6. Set table saw guide to cut through red painted surface, thus exposing white material beneath, usually about 1/8 in. is sufficient to represent brickwork. Use saw which removes about 1/8 in. wide area, and saw crosslines on entire front and side surfaces. Horizontal lines are 1/2 in. on centers, vertical lines are 1 in. on centers.
7. Using regular Styrofoam glue (Rez-N-Glue), seal main body to base, and top panel to top, and candles into position. In order for the candles to fit snugly against side, it is necessary to cut out a circular portion of top piece to allow close fit.
8. Cut two angels to shape.
9. Seal angels to base portion in approximate positions given.
10. Allow to thoroughly dry.

LAMPSHADES FROM POLYPLASTEX

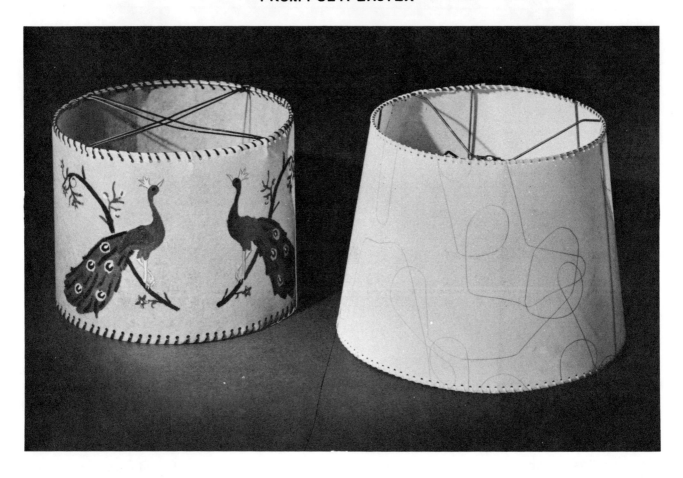

SUBMITTED BY:
Mrs. Nancy McCall
Art Instructor, Hazelwood High
St. Louis, Missouri

With the introduction of the vinyl based Polyplastex lampshade material a fine new field was opened for the "do-it-yourself" fan in making his own lampshades. Almost anyone that can punch a straight line of holes with a paper punch can create an attractive new lampshade, or re-cover an old shade with this easy-to-use material. An array of splendor in literally hundreds of designs and color combinations is available. The material has a fiberglas filler in the case of the "synskin" based designs. The heavier weight stock in the double lamination of the vinyl sheet with decorative center lamination has many many possibilities. Some are laminated with colored threads between, some with real feathers, real leaves and decorative grasses, and even some with pressed butterflies. So the combinations are unlimited. The material is easily cut with scissors and is punched along the edges with a regular round hole paper punch. A special cement is available to make a neat lap joint. The material is laced to the metal frames with regular plastic lacing of a matching or contrasting color. A shade made from plain color material may be painted with your own design (using enamels) such as shown in peacock design photograph.

MATERIAL REQUIRED:
(A) For straight cylinder shape:
1 piece Polyplastex of suitable color 12 in. wide x 48 in. long
1 wire shade frame (can be purchased commercially) or can be made from clothes line wire or clothes hanger wire, curved and soldered to desired size and shape
Approximately 6 yards 3/32 in. wide plas-

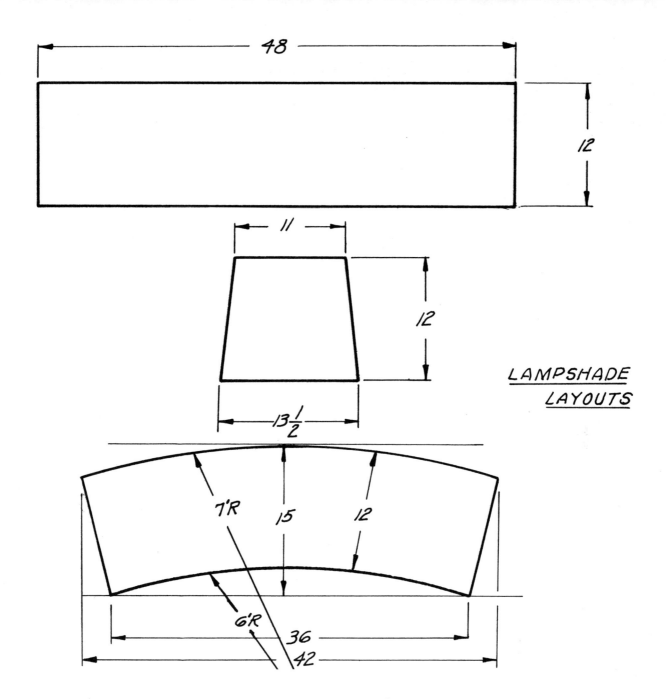

LAMPSHADE LAYOUTS

tic lacing of suitable color
(B) For tapered shade:
1 piece Polyplastex 15 in. x 42 in. of suitable color
Approximately 5 yards of lacing
1 wire shade frame

PROCEDURE:
1. Cut shape of Polyplastex material to shape. Use scissors and make good straight cuts.
2. Approximately 1/4 in. from each long edge, mark centers of holes to be punched, 3/8 in. apart, and punch with 1/8 in. round paper punch. Punch holes evenly and accurately.
3. Start lacing from inside shade frame, and make loops from center of shade out, over the edge and wire, and back under to inside and next hole. Continue around total perimeter and make final loop inside and glue to inside of shade. Repeat around opposite long side.
4. At the joint where the two ends lap over, trim this off so that an even 1/4 in. of lap joint appears, then seal along this joint with Polyplastex cement. Allow to dry thoroughly.

Napkin holder from translucent and opaque plastic.

NAPKIN HOLDER

SUBMITTED BY:
Phil Brooks
University of Missouri
Columbia, Missouri

MATERIAL REQUIRED:
- 2 pieces colored translucent plastic 1/8 x 5-3/8 x 8-3/4 in.
- 1 piece black opaque plastic 3/4 x 3 x 3 in.
- 2 pieces 1/2 in. diameter clear cast plastic rod 2-1/2 in. long

PROCEDURE:
1. Make template for two side plates.
2. Lay out side plate shape on colored plastic.
3. Cut 3 in. x 3 in. black base to size, and bevel at 45 degree angle about half way through thickness.
4. Cut two rods to length (2-1/2 in. long) and sand ends at 10 degree angles.
5. Sand and polish all edges of colored plates and base.
6. Set circular saw blade at 10 degree tilt and saw 1/8 in. wide grooves in black base at location on drawing. Be sure to reverse block so that the angles cut are opposite.
7. Remove all masking papers.
8. Using solvent (Pleximent), seal two colored plastic plates into grooved base and allow to thoroughly dry.
9. Soak ends of rod in solvent, and seal in place as indicated on drawing. Allow to thoroughly dry before handling.
10. Clean and wax.

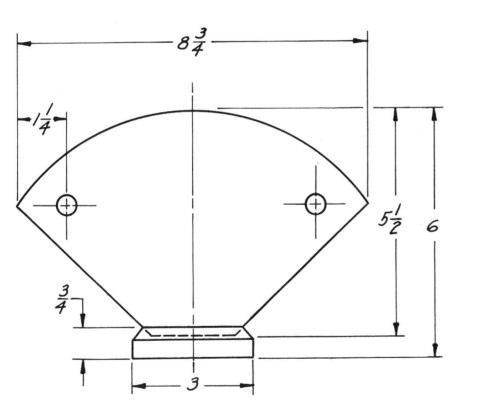
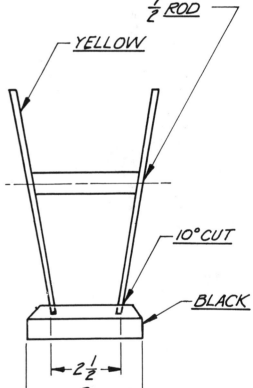

NAPKIN HOLDER

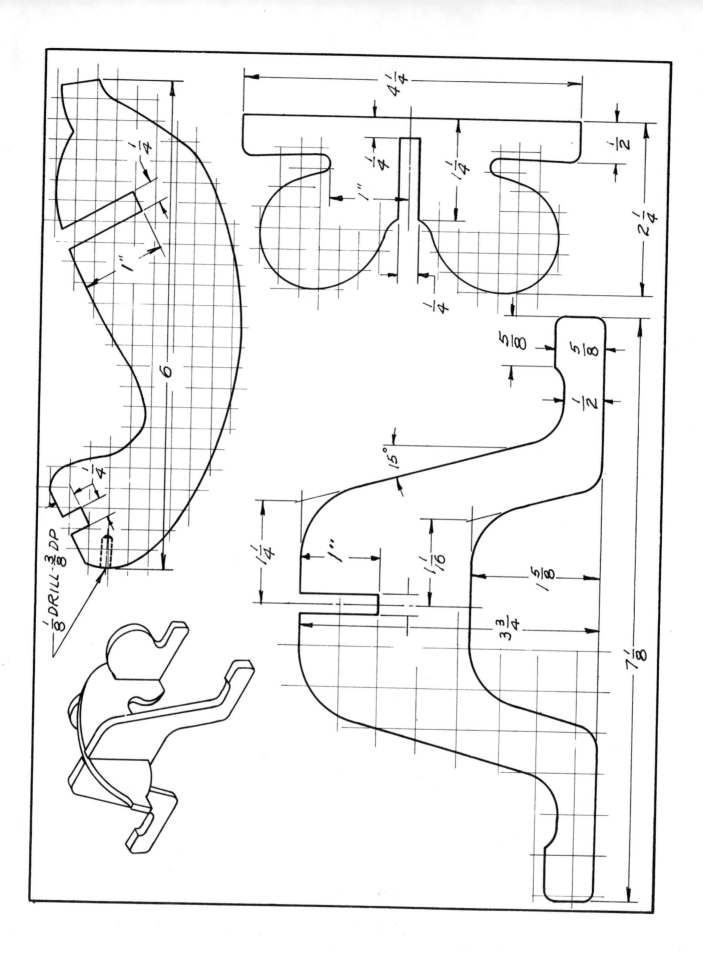

"FIDO" LETTER AND NOTE HOLDER

SUBMITTED BY:
Walter Ambrose
Hadley Technical High School
St. Louis, Missouri

MATERIAL REQUIRED:
1 piece clear plastic 1/4 x 2-1/4 x 4-1/4 in.
1 piece clear plastic 1/4 x 2-3/8 x 6 in.
1 piece clear plastic 1/4 x 3-3/4 x 7-1/8 in.
1 white pipe cleaner (to be cut in two pieces-- one for white eye section and one for tail)
1 wood clothes pin

PROCEDURE:
1. Make templates per drawings for dog parts.
2. Trace templates off onto plastic masking.
3. Cut three piece body to shape. Make sure slots are cut accurately to 1/4 in. to allow slip fit interlock with matching pieces.
4. Sand and polish all edges.
5. Drill 1/8 in. dia. hole 3/8 in. deep into edge of main body section for receiving of pipe cleaner tail.
6. Using approximately a 4 in. pipe cleaner, cut this into a 1 in. piece and a 3 in. piece. Glue the 3 in. piece into drilled hole in main body section and bend to gentle curve to similate upturned tail of dog.
7. Insert 1 in. piece of pipe cleaner through clothes pin hinge, and glue into position.
8. Glue wood clothes pin onto flat area of main body section, and allow to dry thoroughly.
9. Assemble three sections by interlocking matching pieces. Do not glue, unless necessary to hold pieces together. These should be a tight fit, and interlock properly.
10. The feet sections of the two parts may be painted with a colored lacquer to dress up the dog, or a crystalizing colored lacquer may be used. The clothes pin should be coated with colored lacquer or paint to dress it up a little.
11. Clean plastic and wax.

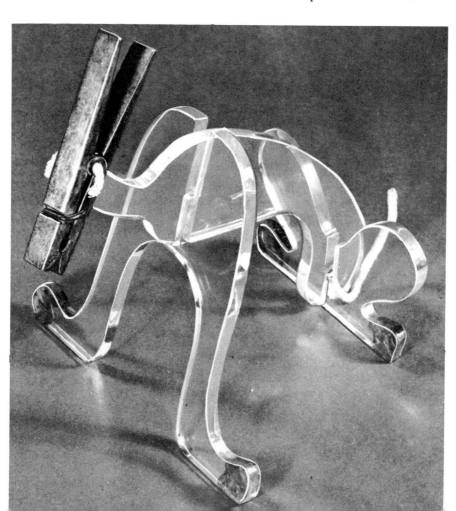

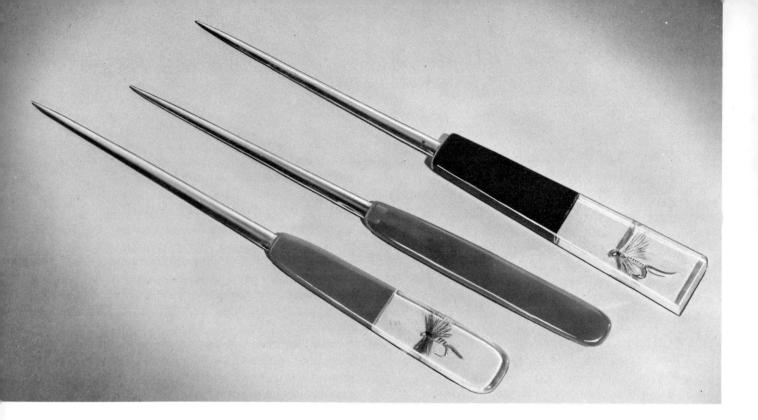

LAMINATED LETTER OPENERS

SUBMITTED BY:
Edison Clark
Sunnyside Jr. High
Lafayette, Indiana

Laminating pieces of various colors of plastics offers unlimited possibilities for turning out novel projects. In making laminated letter openers not only laminating will be described, but also embedding an object such as a fish fly.

Two pieces of plastic will fuse if heated to a temperature of 260-280 degrees F. and pressed together. The temperature will vary with the thickness of the material to a slight degree, but this range can generally be used.

MATERIAL REQUIRED:
(For black and white opener with fish fly)
2 pieces clear plastic 3/16 x 1 x 2-3/8 in.
2 pieces black plastic 3/16 x 1 x 2-3/8 in.
1 letter opener blade (commercially purchased)
1 metal clamp with bolts which are tightened to provide pressure

PROCEDURE:
1. Place two pieces of plastic with the object to be embedded and the letter opener blade in between and tighten bolts of clamp (smooth shank openers should be notched with file). Heat in an oven; upon removal tighten clamps slightly. This method is time consuming as the clamps must heat before the plastic becomes warm enough to fuse.
2. A second method is to put the object to be embedded in between two pieces of plastic. Place on a piece of paper in an oven. Do not place the plastic on metal as sticking and uneven heating of the two pieces of plastic will result in unsatisfactory results. Adjust the temperature of the oven so the two pieces of plastic are heated enough to fuse. Then, quickly remove from oven, and clamp between two wooden blocks with surfaces covered with aluminum. The aluminum provides a smooth surface and prevents the wood grain from marring the plastic. The thicker the object, the thicker the plastic

Projects From Plastics

and the more pressure needed. This method is quicker than the preceding method.

3. A third method requires two electric irons with thermostats such as used in the ironing of clothes. Mount one in a vise or a special bracket in an inverted position. The thermostats are set at slightly above "medium" for 1/8 in. or 3/16 in. plastic used in the embedding of fish flies. The thermostats are set up to almost high for 1/4 in. plastic used in making cake and pie servers. Experimenting will soon reveal the proper temperature.

When the two irons are hot, place the two pieces of plastic on the inverted iron. Let these become hot enough to become flexible like a piece of rubber. Make certain the top surfaces of the plastic are free of lint and adhesive from masking paper. Lay the object to be embedded on one piece. With a pair of of tweezers, or gloved hands, place the other piece of plastic on top. Hold the second hot iron on top of this and press down lightly. If the temperature is correct, the two pieces of plastic will fuse together. By pressing on one corner of the top iron, the plastic should squeeze together rather easily. If the irons are too hot, bubbles will appear in the plastic. Quickly remove the plastic "sandwich" from the irons with a pair of tweezers and apply pressure lightly in a wooden clamp as mentioned previously. The embedded project may be worked as soon as it is cool--usually in two or three minutes.

Difficulty may be encountered with the metal letter opener blade. The metal conducts the heat away around the outer edge and the plastic may not get hot enough to fuse together although it is molded around the blade. This problem is solved quickly. After the plastic has cooled, put several drops of solvent cement at this point and put light pressure on it. If too much pressure is used, the solvent will be squeezed out and may not hold.

Upon completion of the embedment, if an irregular edge results, this should be sanded down on disc sander, then fine sanded, and buffed to high gloss. A little experimenting along these lines will enable you to turn out some beautiful letter openers.

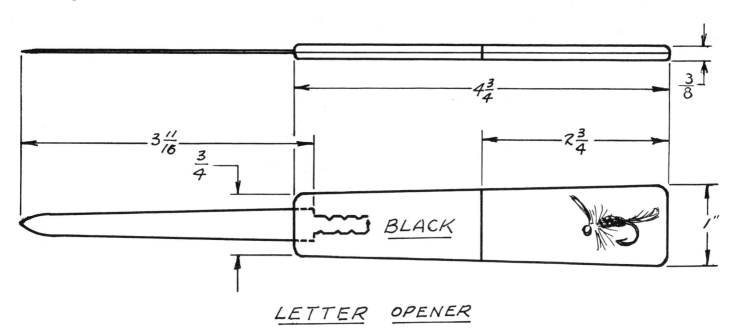

LETTER OPENER

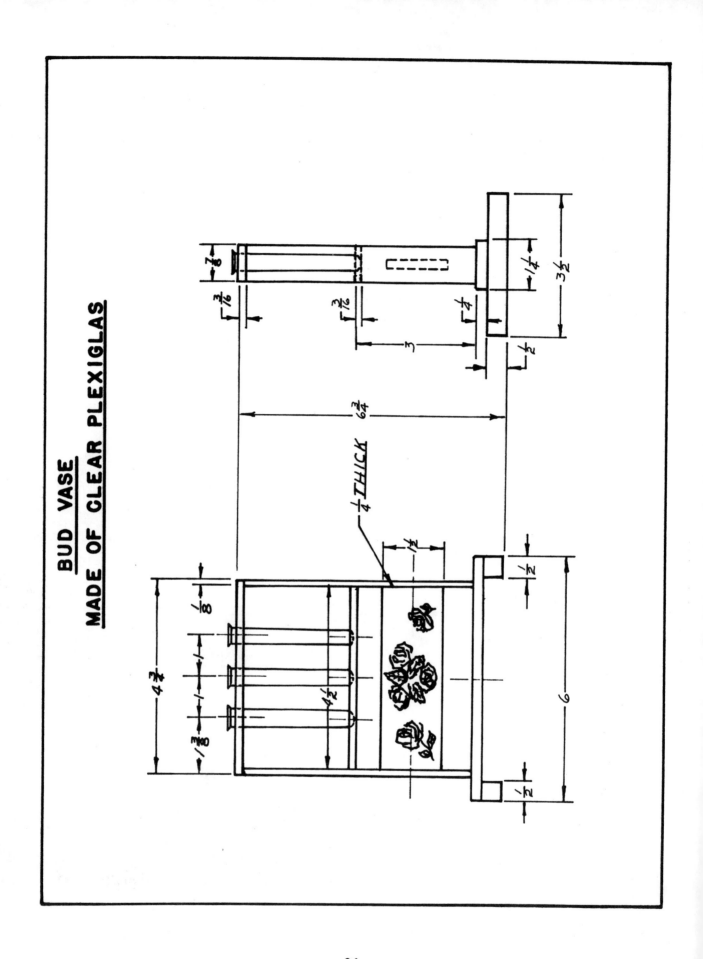

PLEXIGLAS BUD VASE

SUBMITTED BY:
Al Belanger
Master Craftsman
Alton, Illinois

It is difficult to find a good project which can be made from "scraps" of material, and yet finish out into a worthwhile project. Such an item has been developed by Al Belanger, in this unique Plexiglas bud vase. The glass vials are standard size test tubes and can be purchased at drug stores, or obtained from the chemistry department of your school.

MATERIAL REQUIRED:
- 2 pieces 1/8 x 7/8 x 5-3/4 in. clear Plexiglas
- 2 pieces 3/16 x 7/8 x 4-1/2 in. clear Plexiglas
- 1 piece 1/4 x 1-1/2 x 4-1/2 in. clear Plexiglas
- 1 piece 1/4 x 1-1/4 x 6 in. clear Plexiglas
- 2 pieces 1/2 x 1/2 x 3-1/2 in. clear Plexiglas
- 3 small glass test tubes or vials measuring slightly under 1/2 in. diameter

PROCEDURE:
1. Cut all Plexiglas parts to size.
2. Wet sand and polish all edges of all pieces of Plexiglas except the ends of the pieces which form the inside braces and the carved section.
3. Tape the two pieces of Plexiglas 3/16 x 7/8 x 4-1/2 in. together. Drill 1/2 in. diameter holes through one piece, and just spot drill into the second piece as indicated. This taping and drilling together assures alignment after assembly.
4. Internally carve, or surface carve, the 1/4 x 1-1/2 x 4-1/2 in. lower brace as you wish.
5. Solvent-seal sections together as indicated in drawings and photograph.
6. Clean and wax.

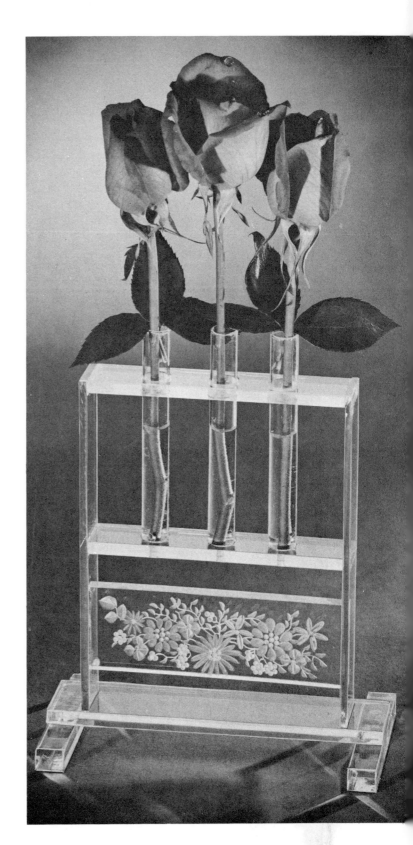

DESK SET With LETTER CLIP

SUBMITTED BY:
David L. Wallace
Craftsman
Alton, Illinois

Usefulness of this desk set is increased by adding a letter clip of simple design.

A small wood jig may be used in heat forming the harp-shaped clip, or the plastic strip may be curved by hand around a tube or rod of suitable size, and held in position until cool.

MATERIAL REQUIRED:
1 piece 1/2 x 4 x 6 in. clear (or colored) Plexiglas
1 piece 1/8 x 3/4 x 8-1/2 in. clear Plexiglas
1 Fountain Pen and Swivel-funnel
1 piece threaded rod (or bolt) 3/8 x 6-32.

PROCEDURE:
1. Saw the clear base block 1/2 x 4 x 6 in To minimize waste, include the sawcut in the 4 x 6 in. measurement, thus finishing to about 3-7/8 x 5-7/8 in.
2. Saw the clear strip 1/8 x 3/4 x 8-1/2 in.
3. Bevel the clear base block about 3/16 in. across at 45°. This can be accomplished on table saw or disc sander.
4. Wet sand edges of base block and 1/8 in. strip.
5. Buff edges to high gloss on cloth buffer.
6. Drill and tap 6-32 hole into base block at indicated point, to 1/4 in. depth. Insert threaded rod, and tighten into threaded hole to bottom, allowing 1/8 in. of rod to protrude.
7. Make wood jig, or use 2 in. diameter cylinder to form letter clip. Heat, form strip to desired shape, and allow to cool.
8. Flatten off small area on formed clip so it makes positive contact with base, then solvent-seal to base at desired point of contact.
9. Thread swivel-funnel onto threaded stud.
10. Clean and wax.

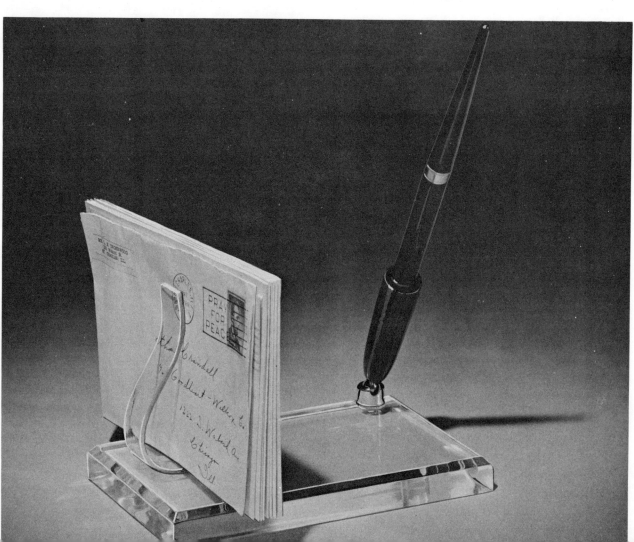

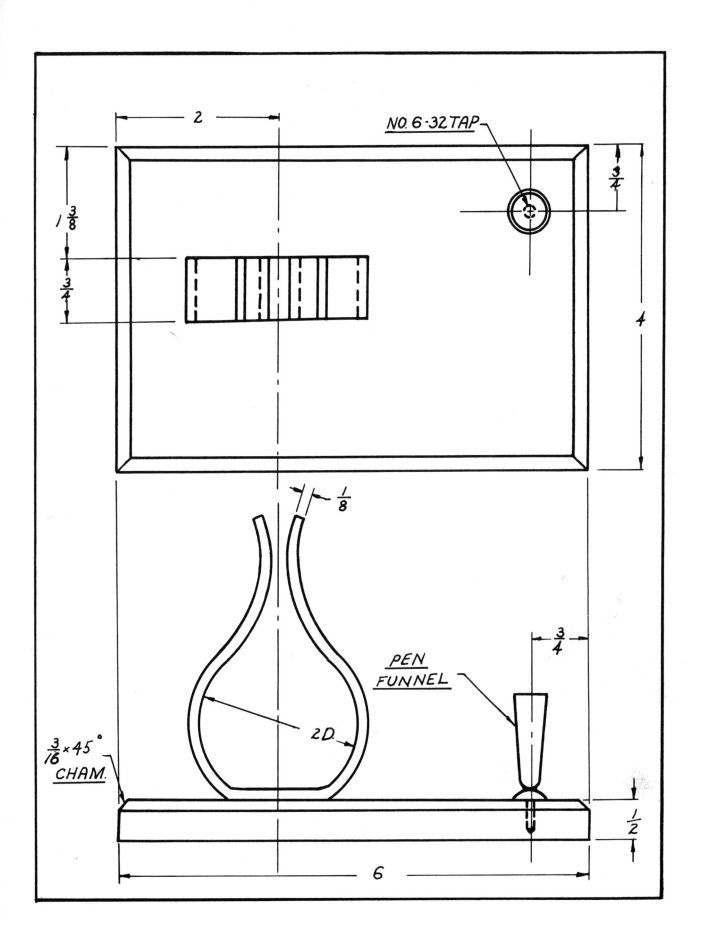

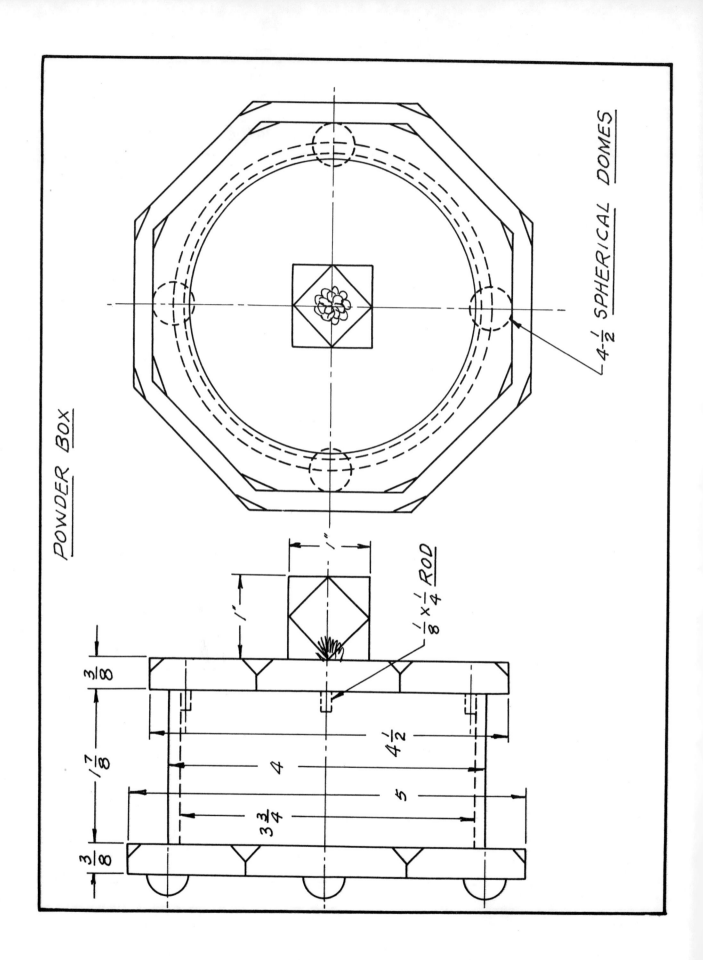

POWDER BOX WITH OCTAGON BASE

SUBMITTED BY:
Clyde Brown
Carthage High School
Carthage, Illinois

This powder box made of clear Lucite tubing with capped end, and matching lid is a very attractive vanity piece, and is very easy to make. Although the project as listed uses clear material throughout, the bottom and top lid can be made of a colored plastic which adds color and beauty to the finished product. This color can be selected to match the decor of the room in which it is to be placed, thus adding a very decorative and personal touch. The top knob can be either a round piece of clear Lucite, or it may be an internally carved item such as the 1 in. faceted cube pictured.

MATERIAL REQUIRED:
1 piece clear cast Lucite tube 4 in. O.D. x 1/8 in. wall x 1-7/8 in. long
1 piece clear plastic 3/8 in. x 4-1/2 in. x 4-1/2 in.
1 piece clear plastic 3/8 in. x 5 in. x 5 in.
1 piece clear plastic 1 in. x 1 in. x 1 in.
4 pieces clear Lucite rod 1/8 in. dia. x 1/4 in. long
4 only #61 clear domes for feet (may be purchased ready-made.

PROCEDURE:
1. Cut 4 in. O.D. tube to length and sand both edges (ends) square.
2. Lay out and cut both top and bottom octagon panels to size using circular saw or band saw. For added beauty of reflection, sand all 8 corner points at 45 degree on both top and bottom.
3. Lay out center points of each side of 1 in. cube, and mark line from center to center of each adjacent side. Sand 1 in. cube to 14 faceted shape.
4. Cut 4 pieces clear Lucite rod 1/8 in. dia. x 1/4 in. long.
5. Sand and polish all edges of all pieces and one end only of each of the four small 1/8 in. dia. x 1/4 in. long rods, and one end only of 4 in. O.D. tube.
6. Remove masking paper from all panels.
7. Internally carve the 1 in. facet cube if desired, and dye and fill with white filler powder.
8. Place sanded edge of tube in shallow tray containing Pleximent and allow edge to soak for about 30 seconds. Then place in centered position on 5 in. octagon base and allow to dry thoroughly.
9. Seal unpolished ends of short rod pieces into position on under side of lid to serve as positioning members between lid and tube section. These positions can easily be predetermined by making a drawing of the inside of the tube on a piece of paper, then laying the top side of the lid next to the paper. Position it correctly, then seal the short rods in position just inside the inner line. As soon as partially dry, place the lid in position over tube, and readjust slightly if needed, and allow to dry.
10. Using soak method, seal the faceted cube into exact center of top of lid, and also seal four domes in position on bottom to act as feet. Allow to thoroughly dry.
11. Clean and wax.

PERFUME VIAL HOLDERS

SUBMITTED BY:
V. S. Fox
Lincoln High School
Cleveland, Ohio

Shown in the photo are eight ways to "dress up" the conventional perfume bottle, the usual dimensions of which are 1-3/4 in. long by approximately 9/16 in. in diameter. The various designs combine different colors, carved and other decorated forms, the possibilities being almost unlimited. All designs have in common the fact that they are made of four identical pieces press-fitted onto the bottle, with a decorative piece attached to the screw cap. Such a holder makes a welcome addition to the decor of the average dressing table.

The data which follows give step-by-step details on how dressing up the vial holder shown in the photo to the right--a typical job, is accomplished.

MATERIAL REQUIRED:
1 piece clear plastic 5-1/4 x 1-3/4 x 1/4 in. (The four equilateral triangles can be cut from this piece without undue waste).
1 piece of clear plastic 3 x 1-1/2 x 1/4 in.

PROCEDURE:
1. Saw out the four equilateral triangles by taking a 60 degree cut off the end of the 5-1/4 in. strip of clear plastic. Make successive 60 degree cuts in opposite directions along the strip.
2. At the exact center of each triangle, drill a hole just slightly less than the diameter of the vial. (Approximately 9/16 in. dia.). (NOTE: To find center of the triangle, draw lines from two of the points to the mid-point of the opposite side. The crossing point of these two lines is the center of the triangle).
3. Cut saw kerfs parallel with each edge of triangle for decoration.
   Make sure that inside edge of the kerf does not cut into edge of center hole. Make kerfs on one face only.
4. Sand and polish all edges of triangular pieces.

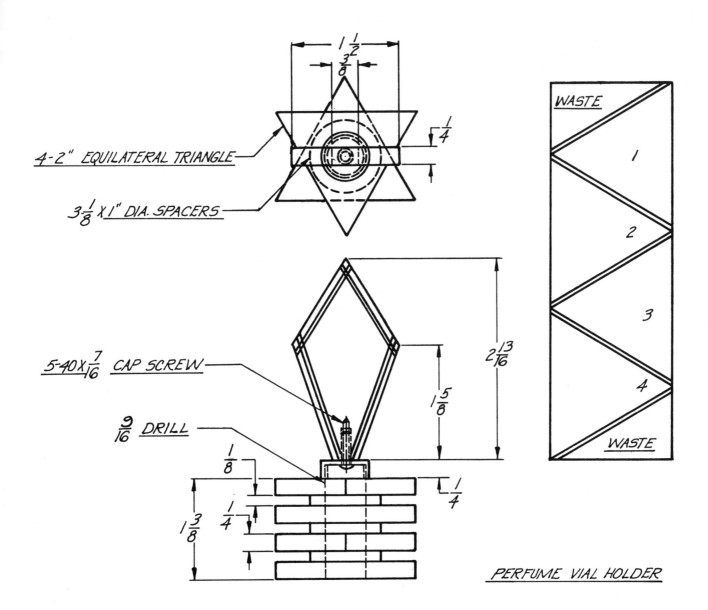

PERFUME VIAL HOLDER

5. Fit over vial by pressing triangles in place, making each one just one-sixth of a turn from the other so that vertices of successive triangles are above mid-point of the side of the one below. Space triangular pieces an equal distance apart (slightly more than 1/8 in.).

NOTE: These vials are not always accurate as to diameter. If triangular-shaped piece will not go down over bottle, carefully heat around edges of hole with Bunsen burner flame or other suitable source of heat. If hole is much too small a ridge may be formed about edge of hole, spoiling effect. Be sure that holes are large enough so that triangles can be press-fitted onto bottle cold, or with only a small amount of stretching when heated.

6. Saw out decorative top piece, leaving a "flat" of slightly less than one-half inch at bottom where piece fits against screw cap of vial.
7. Select machine screw with fine threads. Drill hole of suitable size through top of screw cap. (1/16 in. screw is about right). Drill hole up into top piece just slightly less than outside thread diameter of machine screw. When assembled to cap, machine screw will cut own threads into plastic of top piece.
8. Before attaching top piece to cap, cut decorative saw kerfs along edges of top piece, about 1/8 in. from edge.
9. Polish all edges of top piece and assemble to cap with screw.
10. Wax and polish all parts of piece.

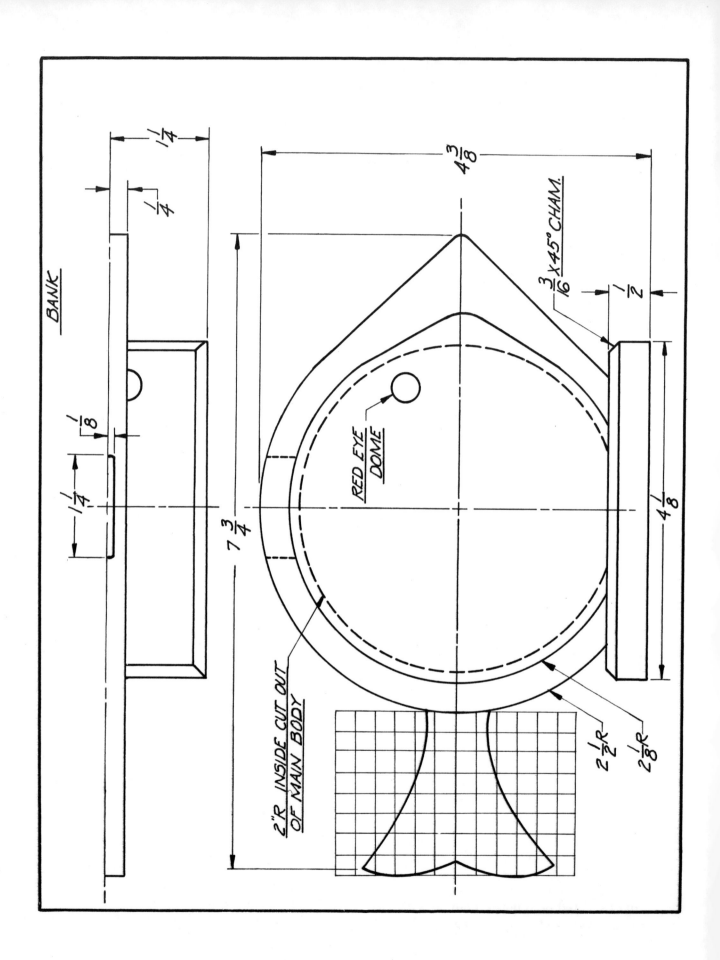

FISH SHAPED COIN BANK

SUBMITTED BY:
Walter Ambrose
Hadley Technical High School
St. Louis, Missouri

Here is a beautiful little project, which, in its basic simplicity, gives a beginner in plastics a chance to fabricate a project he will cherish.

MATERIAL REQUIRED:
1 piece clear or colored plastic 1/2 x 4-3/8 x 7-3/4 in. (main body)
2 pieces white opaque plastic 3/16 in. x 4 x 4-1/2 in. (side plates)
1 piece clear or colored plastic (to match main body) 1/2 x 2-1/2 x 4-1/8 in. (base)
2 only #61 domes, your choice of color, for eyes

PROCEDURE:
1. Cut main body block to size specified, and using compass, locate pivot point, and draw main circular body to 5 in. dia. At the same time, draw inside cutout of body to 4 in. dia. Also draw and complete outside body shape, per drawing.
2. Cut and bevel base at 45 degrees to size, using same color material as main body.
3. Drill hole in stock to be used for main body to take saw blade, saw out as indicated on drawing by dotted line.
4. From 3/16 in. white material, locate pivot point, and draw 4-1/4 in. diameter circle. Lower segment of this will be missing.
5. In top edge of main body locate slot position. Drill two holes 3/16 in. diameter, through edge, entirely through to center cutout section. These holes will form the ends of 1-1/4 in. wide slot for coin entrance.
6. Cut main body of fish to shape.
7. Sand and polish all edges of main body, base block, and two circular side plates-- except flat segment of fish body and side plates to be used for sealing to base.
8. Remove all masking paper.
9. Using soak method, seal side plates to main fish body.
10. Seal #61 eye domes into position, one on each side evenly positioned.
11. After thoroughly dry, set disc sander at 90 degrees and sand bottom of fish body and two side plates to provide perfect flat surface, preparatory to sealing to base.
12. Using soak method, seal fish cutout to base.
13. Clean and wax.

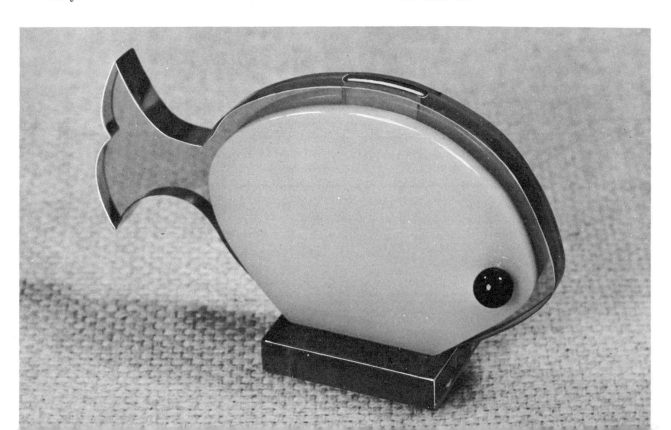

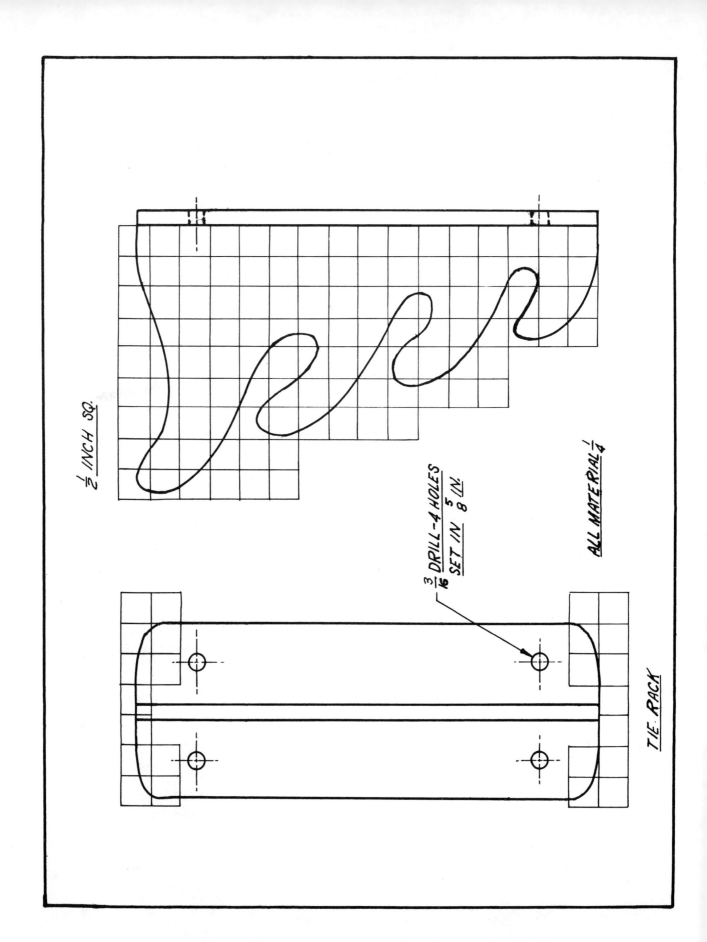

TIE RACK

SUBMITTED BY:
George Melcher
Grant County Rural High School
Ulysses, Kansas

This tie rack made of gleaming black opaque plastic will make a beautiful addition to any man's accessories. The rack may be used for ties, or to hold several clothes hangers. It is sturdy, and easy to make, yet its simple flowing lines have a beauty that is surprising. Making these tie racks provides an excellent exercise for students in sawing basic curved lines in plastics.

MATERIAL REQUIRED:
1 piece 1/4 in. plastic, black, opaque, 7-5/8 x 2-15/16 in.
1 piece 1/4 in. plastic, black, opaque, 7-5/8 x 4-1/2 in.

PROCEDURE:
1. Saw out the projecting piece from the 4-1/2 in. wide piece of plastic, using a scroll saw or band saw.
2. File smooth the edges of the cuts where they are indented, using a rat-tail file to remove all traces of saw marks.
3. Water-sand the straight back edge of this piece to make sure it is straight and square for cementing to back piece.
4. Saw out rectangular back piece.
5. Round off corners and slightly round top and bottom edges, using sanding belt if available. If not, use a piece of abrasive paper laid on a flat surface.
6. Polish and buff all edges.
7. Cement projecting piece to exact center line of back piece, as indicated in drawing. Since this bond must carry the load of weight on rack, be sure that the best possible bond is obtained.
8. Drill four holes, as indicated, 3/16 in. in diameter to take screws used to fasten rack on wall or door.

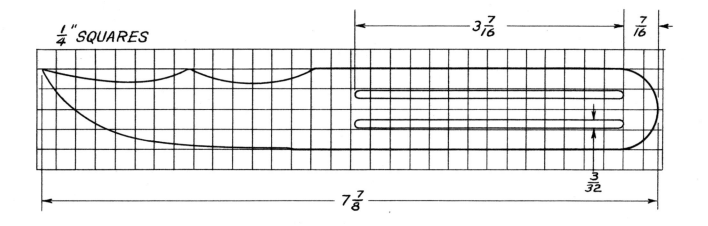

LETTER OPENER AFTER PLAITING

LETTER OPENER

SUBMITTED BY:
Dr. V. N. Hukill
Arkansas State Teachers College
Conway, Arkansas

The unique feature of this project is the method in which the handle is formed, called the endless plait, a trick borrowed from the leatherworking area. Without this treatment, it is just another letter opener. It works well with any color plastic. The design on the blade is optional and may be left to the creative ability of the student.

MATERIAL REQUIRED:
1 piece 1/8 in. thick plastic.

PROCEDURE:
The following procedure indicates how to plait the handle of the letter opener, after it has been heated until flexible.
1. With the blade end up and out, letter the three parts A, B, and C, and the bottom end D.
2. Place C over B and under A. (As in regular plaiting.)
3. Grasp AC in the left hand and fold end D up and through to the right, between A and B. It will look messy and impossible at this point, but be patient.
4. Bring B over A and under C, much as in Step 2.
5. Grasp AC in the right hand and fold end D up and through to the left between B and C.
6. At this point it may need to be reheated for final shaping and straightening.

The plastic must be extremely soft and flexible, and you must work fast. It is suggested that you practice on a piece of leather until you are thoroughly familiar with the steps of procedure. It may be necessary to only do steps 2 and 3, then reheat before proceeding with steps 4 and 5.

By placing the blade end of the opener between two pieces of wood or asbestos in the heating process, it will remain flat and not be distorted in any way.

The two slits in the handle may be made either on the jigsaw or on the drill press. By using a 3/32 in. bit, and exposing only about 3/16 in. below the chuck, the slots may be cut using the bit as a router. However, one must be careful and move the plastic extremely slow.

PLASTIC CUP
With COVER

SUBMITTED BY:
George Melcher
Grant County Rural High School
Ulysses, Kansas

The combination of transparent red, white, and clear plastics used in this project enables the beginning student to visualize the possibilities of color combinations with this versatile material.

MATERIAL REQUIRED:
1 piece clear plastic tube, 3 in. diameter x 2-3/4 in. high, with 1/8 in. thick wall
2 pieces 1/8 in. thick transparent red plastic 3 x 3 in. square.
1 piece 1/8 in. thick transparent red plastic 2-3/4 in. square
1 strip transparent red plastic 7/16 x 3-3/4 x 1/8 in. thick
1 strip opaque white plastic 7/16 in. x 3 x 1/8 in.

PROCEDURE:
1. Be sure that the piece of clear plastic tubing that forms the round body of the cup is sawed off square on both ends. Water-sand the end that is to form the bottom. Sand and buff the top edge of the tubular piece.
2. Sand and polish the edges of the two handle strips. The white strip is to be used as the handle for the lid. The red strip is to be used for the handle for the cup. These pieces should be heated in the heat-controlled oven at about 275 degrees F. until soft enough to bend. They may be heated while other work on the project is progressing.
3. Saw the three circles that compose the lid and the bottom of the cup. The bottom is made from a 3-in. circle of red plastic, 1/8 in. thick. The lid is made from two circles, one 3 in. in diameter, and one 2-3/4 in. in diameter. This smaller circle should be of a size to fit comfortably inside the top of the tubular piece. Smooth and buff edges of all circular pieces.
4. Form the lid by cementing the small 2-3/4 in. circle in the center of the larger 3 in. circle of red plastic.
5. Cement the other 3 in. circle to the sanded edge of the tubular piece to form the bottom of the cup. Be sure the bottom fits tight against the tubular piece all the way around. Cement carefully, and with enough pressure to assure that the bond between the bottom circle and the tubular side is perfect all the way round. This must be done if the cup is to be watertight.
6. Bend the white lid handle strip as shown. When this has cooled, sand the two ends on the sanding belt until they will sit square on a flat surface. Soak ends of this strip in cement and fasten to circular lid, being sure that the handle is cemented to the top, or larger, circle, in the center.
7. Bend the side handle strip of red plastic to the approximate shape shown. When cool, sand the two ends of the handle strip on the sanding belt until they sit square on a flat surface. In cementing this handle to the side of the cup, soak the ends of the handle piece in the cement somewhat longer than usual so that the ends will be softened enough to allow them to conform somewhat to the rounded surface of the side of the cup.
8. Buff and wax all surfaces and edges.

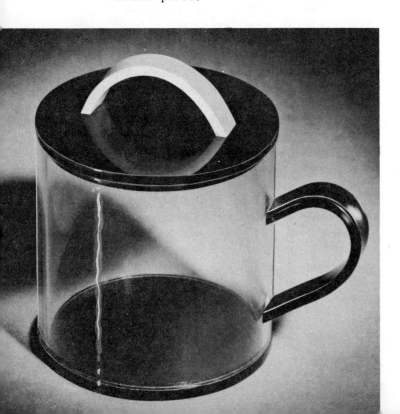

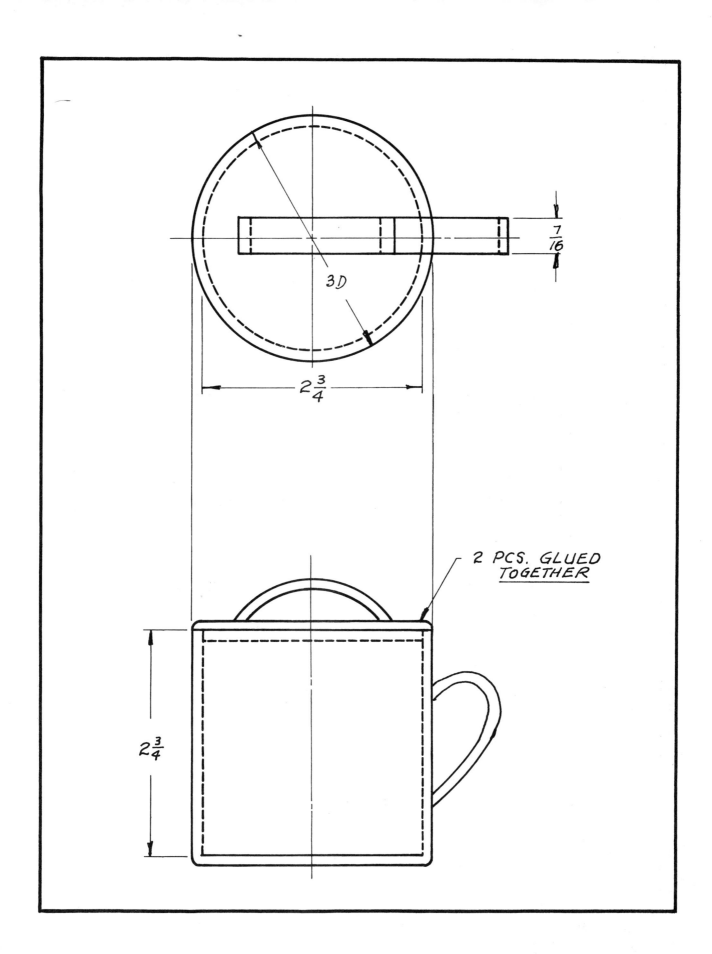

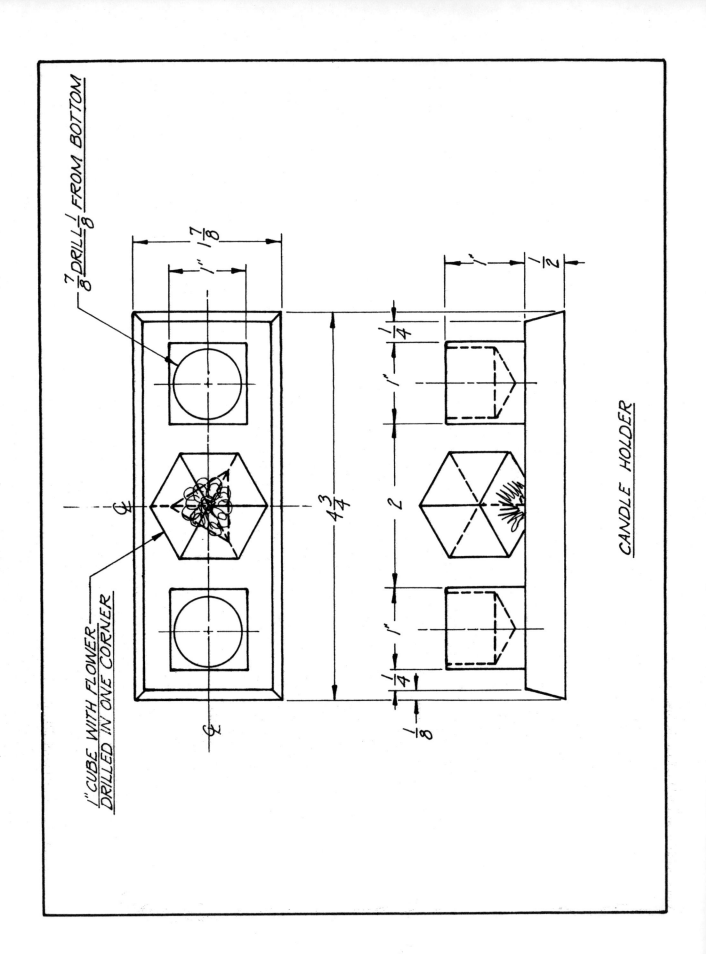

CANDLE HOLDER

SUBMITTED BY:
John Doing
Pacific High School
Pacific, Missouri

In making this project of simple, but attractive design, clear plastic is used for all of the parts.

MATERIAL REQUIRED:
1 piece 1/2 x 1-7/8 x 4-3/4 in. clear plastic
3 pieces 1 x 1 x 1 in. clear plastic

PROCEDURE:
1. Cut base to size 1/2 in. x 1-7/8 in. x 4-3/4 in. and bevel at approximately 20 degree angle on saw, or on disc sander.
2. Cut the three clear plastic cubes 1 in. x 1 in. x 1 in. to size.
3. Lay out diagonal cut on one of the 1 in. cubes so one corner will be cut away and approximately 3/8 in. of the edge is remaining on each of three sides adjacent.
4. Cut corner away to line, and sand flat.
5. Lay out center of two remaining cubes, and drill 7/8 in. diameter hole 7/8 in. deep at center point.
6. Pour solvent (Pleximent) into these two holes, allow to set for about fifteen seconds, then pour out, and allow to thoroughly dry.
7. Sand and polish all edges of base and cubes on all sides except triangular base cut on one cube. Allow this to remain smooth sanded only. Remove masking papers.
8. Internally carve desired design in triangular section of 1 in. cube, dye and fill.
9. Using soak method, seal the two candle holder cubes to position indicated in drawing. Seal internally carved cube to center of base. Allow to thoroughly dry.
10. Clean and wax.

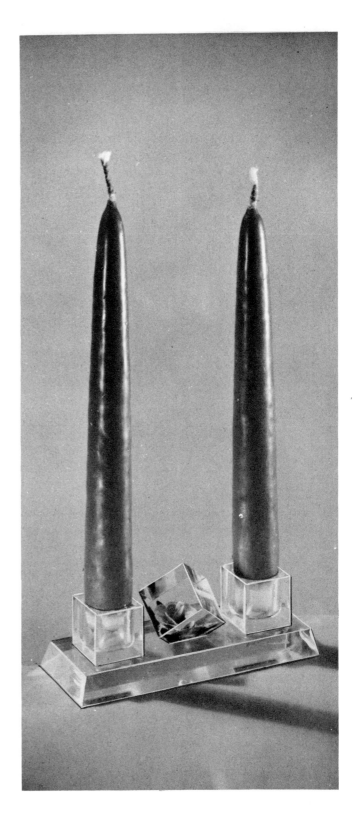

CANDY DISH From
PATTERNED SURFACE PLASTIC

A property that is constantly being explored with plastic, is the heating and forming of the material to various shapes. An interesting variation is the use of plastic with patterned surface. This candy dish is a good example of the attractive projects that can be made with patterned plastic. Patterned surface can be provided by the student or plastic may be obtained with patterns already on the plastic.

MATERIAL REQUIRED:
1 piece clear plastic 3/16 x 8 x 8 in.
1 piece clear plastic 3/16 x 1 x 16 in.

PROCEDURE:
1. Cut two panels to size indicated.
2. Remove masking paper from panels.
3. <u>Preparation of Embossed or Patterned Surface:</u> Glass with pattern on surface or panel of decorative metal such as a meshed grille work may be used in forming pattern on smooth surface plastic.
4. Place panel of plastic in oven at about 275 F. for approximately five minutes. When thoroughly softened, place one side of plastic panel against patterned surface of metal or glass. Place panel of flat glass against the other side.
5. Press the two together with fairly heavy pressure, and hold in position about five minutes. When cool, release pressure.
6. One surface will take the exact shape of pattern. The remaining surface will be smooth like the smooth glass surface.
7. After the patterned surface is obtained, we still must form the panel and handle to shape. This can be done by heating to about 275 degrees F. again, but only until soft enough to form. Prolonged exposure to the heat will remove the surface patterns and the panel will again be flat surfaced.

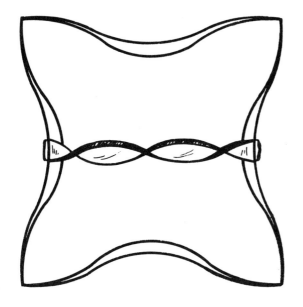

DISH MADE FROM
8" SQUARE OF
PLEXIGLAS

HANDLE MADE FROM
1"X 16" STRIP OF
PLEXIGLAS

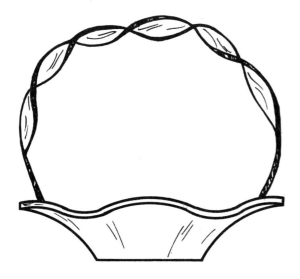

BASKET CANDY DISH

8. When the material is softened enough to form, remove from oven and bend form to approximate shape shown in the photo. Hold in position until cooled.
9. The handle is formed in the same manner, however, it is heated and twisted from 4 to 6 times, and at the same time, formed into arc desired, and held to same width at ends as the width of the tray at its pinched sides.
10. Seal ends of handle to tray as indicated.
11. Clean and wax.

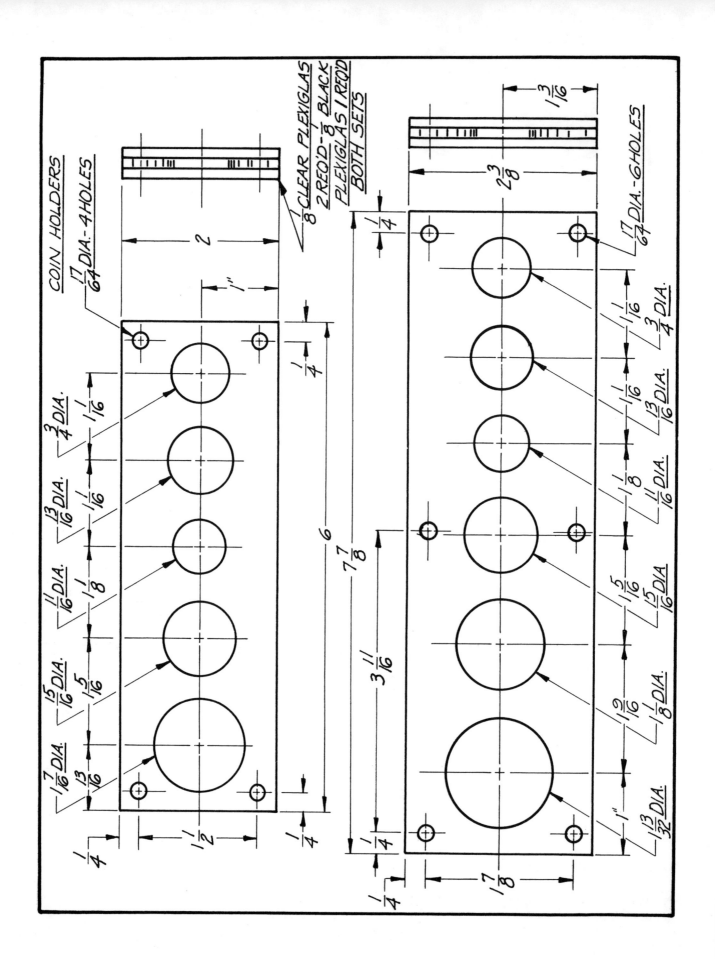

COIN HOLDERS

SUBMITTED BY:
Hazelwood Brothers
Salina, Kansas

A collector of coins can use to good advantage a holder which enables him to view both sides of each coin and at the same time, not handle the coins themselves. The holder shown is comprised of three layers of plastic. The inner core is made of colored plastic, and the two outer panels of clear plastic. The plastic sheets are fastened together with small binder posts, thus permitting the unit to be assembled or disassembled at will. Sketched are the proper sizes and dimensions needed to make up holders for U.S. coins (small holder) and the Canadian coins (large holder).

MATERIAL REQUIRED:
For small U.S. coin holder:
4 binder posts 1/4 in. dia. x 3/8 in. long
2 pieces clear plastic 1/8 x 2 x 6 in.
1 piece black plastic 1/8 x 2 x 6 in.
For Canadian coin holder:
2 pieces clear plastic 1/8 x 2-3/8 x 7-3/4 in.
1 piece black plastic 1/8 x 2-3/8 x 7-3/4 in.
6 binder posts 1/4 in. dia. x 3/8 in. long

PROCEDURE: (For either holder)
1. Cut all three panels required to size.
2. Lay out center line (horizontal) and locate hole positions of various diameters required. This is done on one panel only, the colored center panel.
3. Using clear Cellophane tape, tape the three panels together and lay out and drill holes through all three at one time for binder posts insertion. This hole should be 17/64 in. diameter, thus a slip fit for the 1/4 in. binder post.
4. Drill required holes for coins in the center panel as indicated.
5. Place all three panels together, insert binder posts, and sand edges.
6. Sand polish all edges while together.
7. Remove binder posts, and remove masking paper from all panels. Identity of each coin hole may be made on center panel by gold stamping with hot steel stamp die and gold foil, or may be etched thereon with tiny ball cutter in power tool.
8. Clean all panels, and wax.
9. Reassemble panels with coins inserted, and use binder posts to hold together.

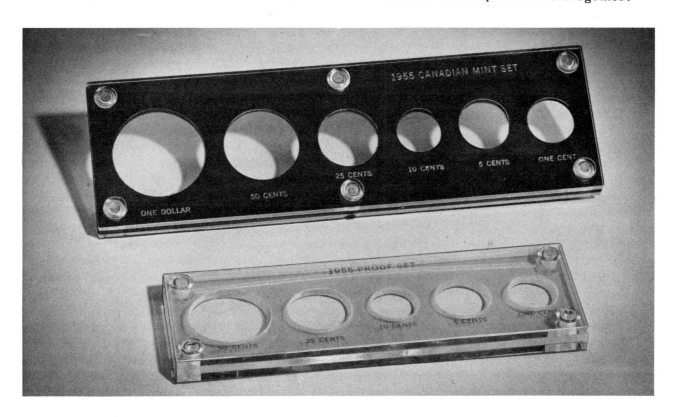

BUD VASE

SUBMITTED BY:
V. S. Fox
Lincoln High School
Cleveland, Ohio

MATERIAL REQUIRED:
1 piece clear plastic 1/4 x 4 x 4 in.
1 piece clear plastic 1/8 x 4 x 8-3/4 in.
1 glass test tube about 3/4 in. dia. x 5 or 6 in. long

PROCEDURE:
1. Test tube should be used to form base of the plastic leaf for fairly close fit, thus assuring proper holding of the tube in place.
2. Cut a piece of plastic 4 x 4 x 1/4 in. (may be clear or colored as you choose). If clear is used, this may be surface carved with floral spray or leaf or fern designs. Bevel and polish base, beveling base at 45 degrees about one-half way through.
3. Sketch leaf design onto masking paper of plastic piece 1/8 in. x 4 in. x 8-3/4 in. Cut to shape on band saw, sand edges and buff.
4. Remove masking paper from both pieces of plastic.
5. Leaf panel is then heated in oven and formed around test tube and flared slightly to pleasing curvature, and held in position until cool.
6. Touch curved base of leaf flat up to disc sander, or sand flat with abrasive paper laid on flat metal surface, to assure smooth surface for next operation.
7. Seal to base section using soak method with Pleximent.
8. Clean and wax.

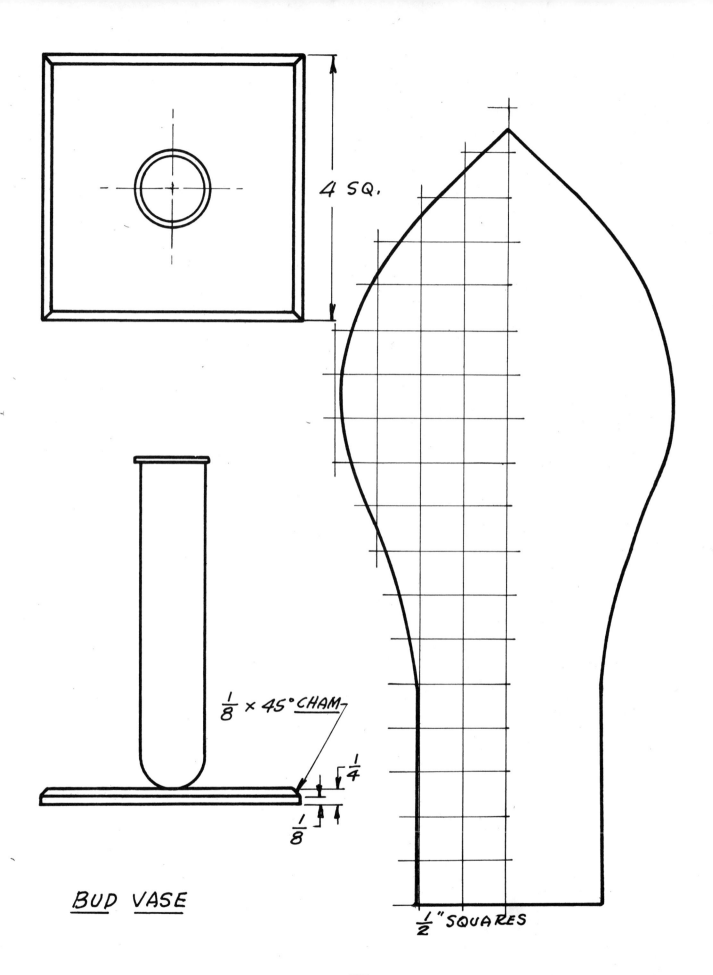

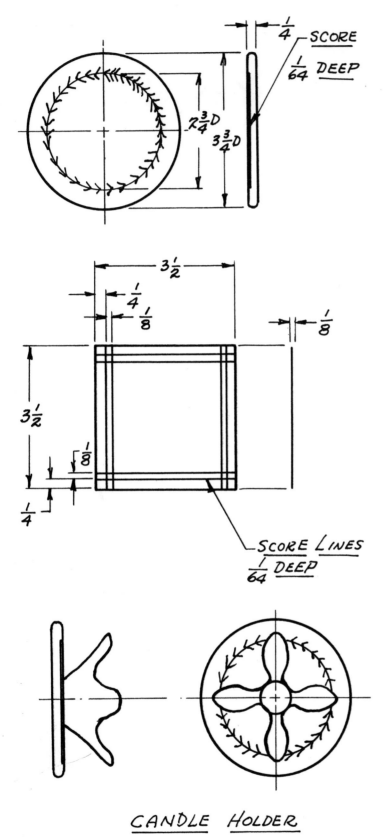

CANDLE HOLDER

FLUTED CANDLE HOLDERS

SUBMITTED BY:
V. S. Fox
Lincoln High School
Cleveland, Ohio

The candle holders shown in the photo consist of clear plastic bases decorated with simple surface carving, and fluted candle receptacles made of red plastic.

The clear bases are decorated by a simple arrangement of drilled holes, similar to those used in internal carving, which makes a leaf "wreath" of design. The fluted red top piece is decorated with parallel scorings cut with the circular saw.

The overall effect is that of simplicity yet colorful decorativeness.

MATERIAL REQUIRED: (For one pair)
2 pieces of clear plastic 1/4 x 3-7/8 x 3-7/8 in.
2 pieces red plastic 1/8 x 3-1/2 x 3-1/2 in.

PROCEDURE:
1. From the square of clear plastic cut a circle approximately 3-3/4 in. in diameter. Scribe the pattern for the circle with a pencil compass on the masking paper and saw out with jig saw or band saw.
2. Smooth the edges of circle until perfectly round, using file, sharp scraper, or other suitable tool.
3. Wet-sand and polish edge of circular base piece, taking care not to round off edge except to remove sharpness.
4. Scribe circle on under side of clear base. Make this scribed circle about 1/2-inch in from outer edge of base.
5. Use small drill to form leaf-like design on both sides of this scribed circle. Handle drill as described in section on internal carving. Drill should enter plastic at approximately 45 degree angle, and should not penetrate more than half-way through the quarter-inch plastic sheet.
6. Remove masking paper from square of red plastic and clean both faces.
7. Score decorative parallel lines on one

Fluted candle holders.

side of sheet. This is best done by lowering table saw until below surface of saw work table, then cautiously raising it until saw teeth just project above table surface. (Try on piece of scrap plastic until you get exactly correct depth of cut). In piece shown in photo, first line is scribed 1/4 in. in from edge of plastic sheet. Second line 1/8 in. in from first.

8. Sand and polish edges of red squares.
9. Heat each red square in oven until soft enough to work.
10. Using gloves, bend sheet up in flutes as shown, using a 3/4 in. wooden dowel pin as forming block. Press pin firmly against center of sheet and lift up flexible hot sheet at center point of each edge. Hold in place until cool.
11. Cement fluted red holder portion to base.
12. Clean completed holder and wax.

INTERNALLY CARVED PIN AND EARRING SET

SUBMITTED BY:
Oliver Oberlander
Southeastern State College
Durant, Oklahoma

Pin and Earring Sets make very interesting projects. In this set internal carving was done in the 3/16 in. thick clear top part. This was backed up with colored 1/16 in. plastic (light green). It might be well to mention that the 3/16 in. thickness, normally considered quite thin for carving, was used in this case as the carving was kept shallow. Plastic 1/4 in. can be used if desired.

MATERIAL REQUIRED:
- 2 pieces clear plastic (carvable) 1 x 1 x 3/16 in.
- 1 piece clear plastic (carvable) 1-5/8 x 1-5/8 x 3/16 in.
- 2 pieces light green plastic 1 x 1 x 1/16 in.
- 1 piece light green plastic 1-5/8 x 1-5/8 x 1/16 in.
- 1 only 1 in. nickel pinback and 2 nickel ear-screws

PROCEDURE:
1. Cut above listed pieces to size on table saw.
2. Remove masking paper from all parts.
3. Internally carve appropriate design in the three pieces of clear stock.
4. Dye (or leave white if desired) and fill with white filling plaster. Allow to dry hard.
5. Lightly sand the surface into which carving is done, in order to have good flat surface for sealing. NOTE: Plaster should fill in the carved cavity completely, otherwise solvent used for next operation will seep into carving and ruin dye.
6. Using the green 1/16 in. stock, soak the surface of one side in Pleximent (ethylene dichloride) for approximately one minute then place in position on matching size of clear carved block. Press firmly and evenly, but do not apply clamping or extreme pressure. Allow to solvent weld and to dry thoroughly.
7. After thoroughly dry, shape pin to pattern as shown or other suitable shape.
8. Set bevel at 45 degree (on disc sander) and bevel all four sides of each so that about 1/16 in. flat remains on edge.
9. Wet-sand smooth, buff and polish to high luster.
10. Cement pinback in place on pin, and pair of earrings in place on the earring blanks. Allow to dry. Clean and wax.

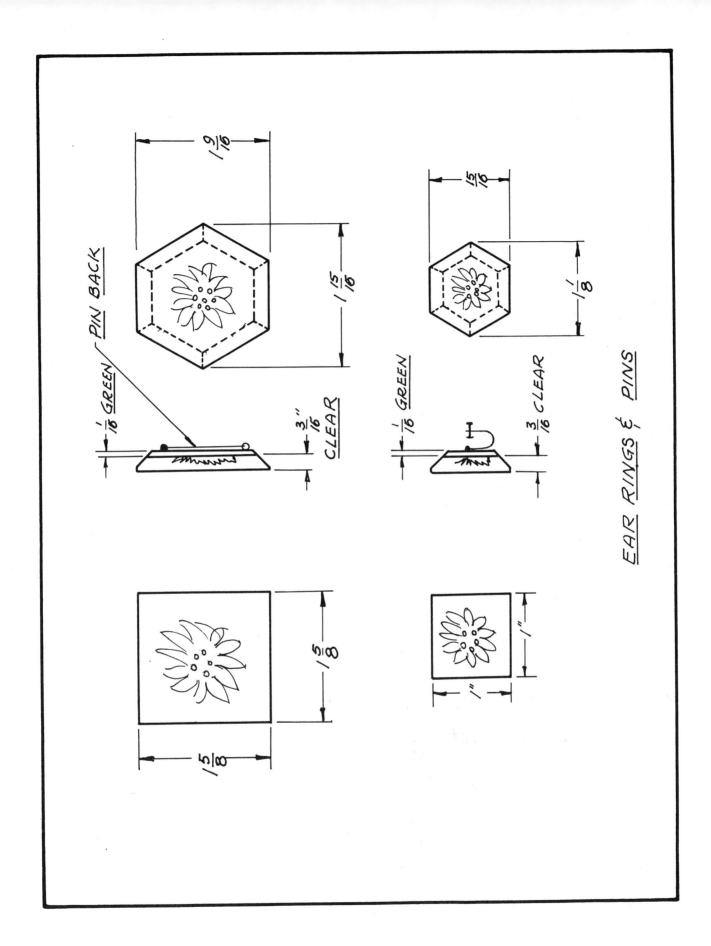

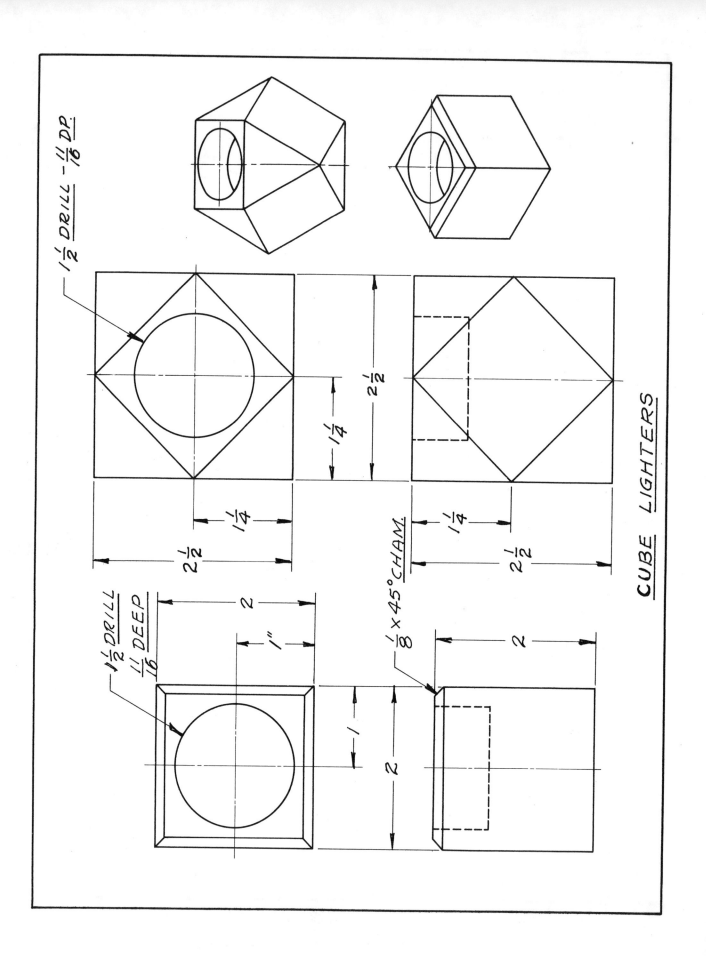

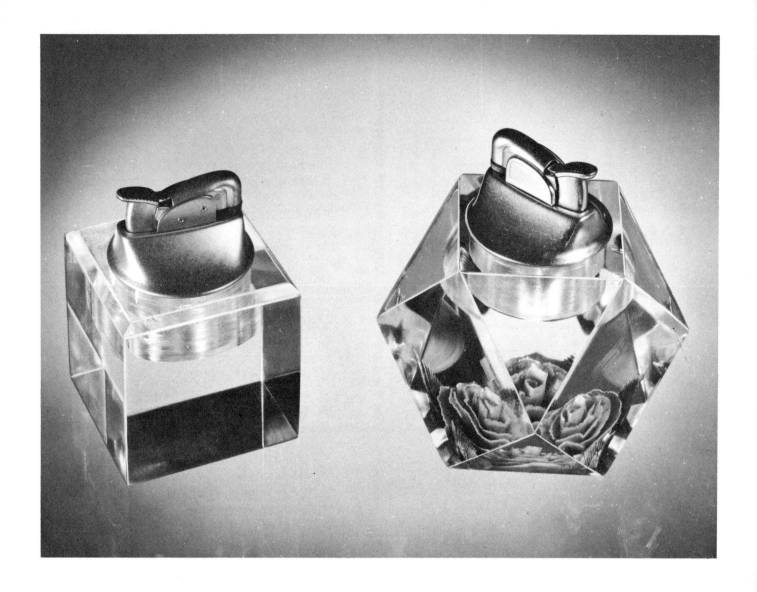

CUBE LIGHTERS

This project uses either 2 in. or 2-1/2 in. cubes for mounts and can be varied to suite the craftsman. Internal carving is shown on one of the blocks, but this is optional. Just the clear crystal-like polished plastic is very beautiful with either gold or silver finished lighter units, and in many cases is left just that way.

MATERIAL REQUIRED:
  1 only 2 in. clear plastic cube
  or 1 only 2-1/2 in. clear plastic cube
  1 Evans lighter unit (1-1/2 in. short model)

PROCEDURE:
1. Cut cube to size. In the case of the 2 in. cube bevel as indicated in photo (45 deg.).
2. If 2-1/2 in. facet cube is used, cut to cube first, then mark centers of all edges with pencil. Draw diagonal line from center to center of adjoining edges.
3. Make diagonal cuts on band saw about 1/16 in. away from lines drawn, thus cutting all eight corners of the cube.
4. Sand all edges of cube, either the 2 in. straight cube or the 2-1/2 in. faceted cube.
5. Mark center of cube, then bore 1-1/2 in. diameter hole 11/16 in. deep to receive lighter unit. Solvent polish hole if desired.
6. Polish all edges.
7. Clean and wax.
8. Insert lighter unit in block.

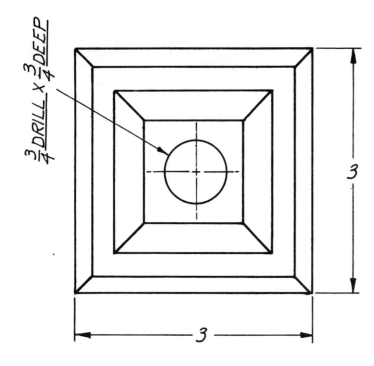

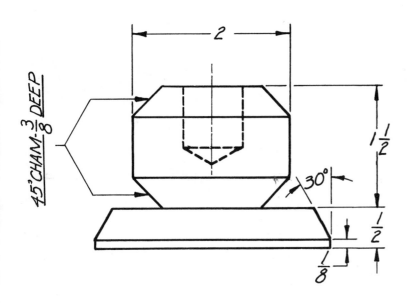

CANDLE HOLDERS

PAIR of CANDLE HOLDERS

SUBMITTED BY:
Lloyd Gouine
Cheboygan High School
Cheboygan, Michigan

Nothing is richer in beauty than highly polished, sparkling plastic in its crystal clarity. We present this pair of beveled candle holders which reflect this beauty from every facet.

MATERIAL REQUIRED: (For one pair)
2 pieces clear plastic 1/2 x 3 x 3 in.
2 pieces clear plastic 1-1/2 x 2 x 2 in.

PROCEDURE:
1. Cut the four pieces of material as described above.
2. Set saw, or tilt disc sander, to 30 degree angle, and either saw bevel or sand bevel the 3 in. base pieces.
3. Reset saw, or tilt disc sander, to 45 degree angle, and saw or sand bevel on both faces of 2 in. x 2 in. block.
4. Lay out location of hole to be drilled in 1-1/2 in. thick block.
5. Drill hole, 3/4 in. diameter, 3/4 in. deep.
6. Sand and polish all edges.
7. Remove all masking paper.
8. Solvent polish hole by pouring solvent (Pleximent) into hole, allowing to remain therein for about 15 seconds, then pouring out again. Allow to dry.
9. Using the 1-1/2 in. thick block, place flat surface (one opposite hole) on tray and fasten with Pleximent; soak, and seal center to base section. Dry thoroughly.
10. Clean and wax.
11. Insert candles of your color choice into holes.

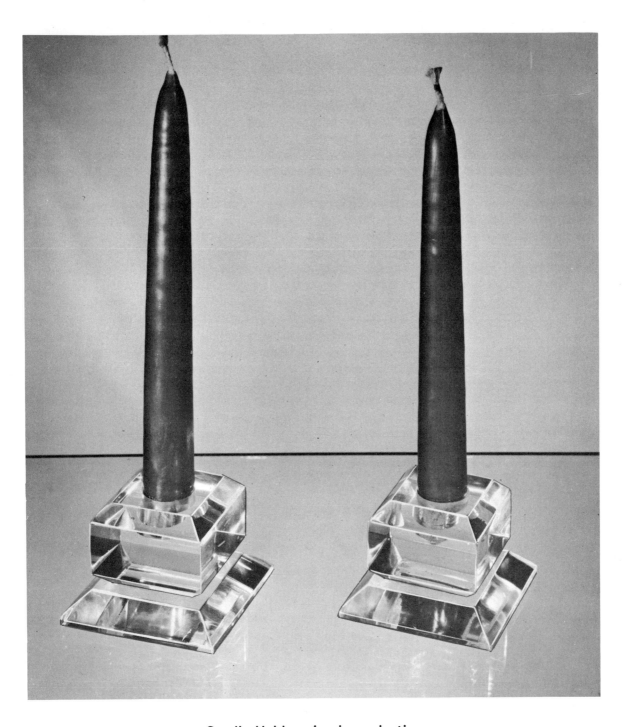

Candle Holders in clear plastic.

ACRYLITE MAGAZINE RACK

One of the most decorative and beautiful plastic materials to come to the foreground is the material known as "Acrylite." We used this material in the Abaca or straw pattern in making the magazine rack of modern design, shown in the photo. Acrylite is easy to work, and is very similar to Plexiglas in all its characteristics. You'll enjoy working with Acrylite, and will be enthused over the decorative beauty of the finished projects.

MATERIAL REQUIRED:

1 piece ST-1 Abaca (straw) Acrylite (or any other suitable pattern), 1/8 x 12 x 24 in.

2 pieces 1/4 x 9 x 10 in. #2025 black Plexiglas

2 pieces 1/2 x 1/2 x 9 in. #2025 black Plexiglas square rod

PROCEDURE:
1. Cut Acrylite to size and the two square Plexiglas bars to length.
2. Lay out pattern for two base holder sides, and cut holders to shape on band or scroll saw.
3. Wet sand and polish edges of all parts.
4. Using a strip heater, form the sheet of Acrylite at given lines, and allow to cool. These may be formed to fit well into the side holder section previously cut.
5. Solvent-weld the two square crossbars to the two side panel holders, allowing to dry thoroughly. Be sure bottom edges rest on same plane by letting them dry in this position.
6. Place Acrylite rack into position. This may be sealed in place if desired, or left as a two-piece unit.
7. Clean and wax.

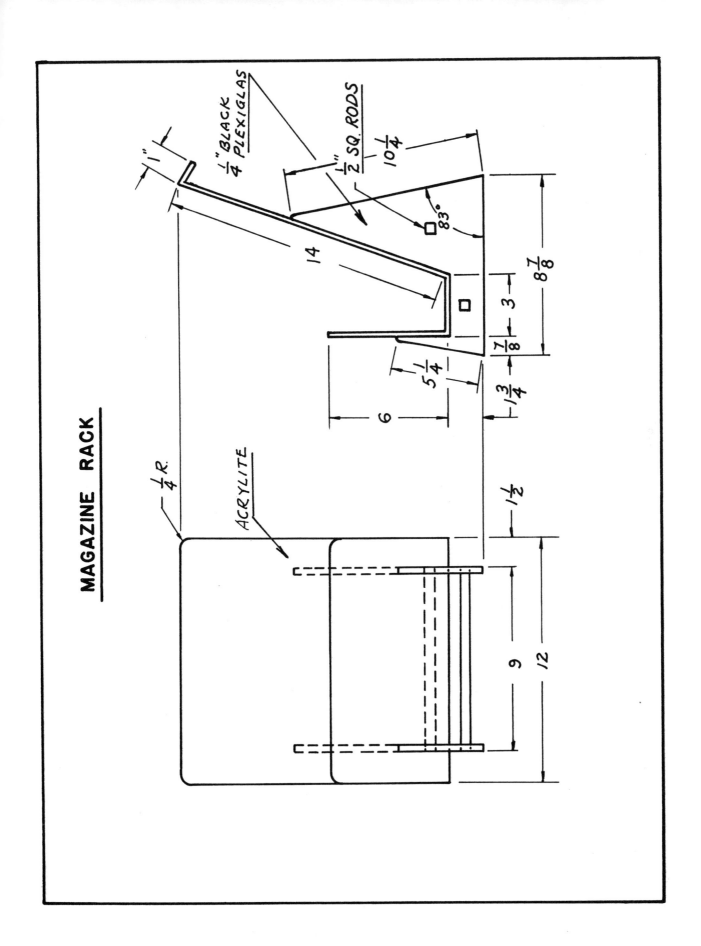

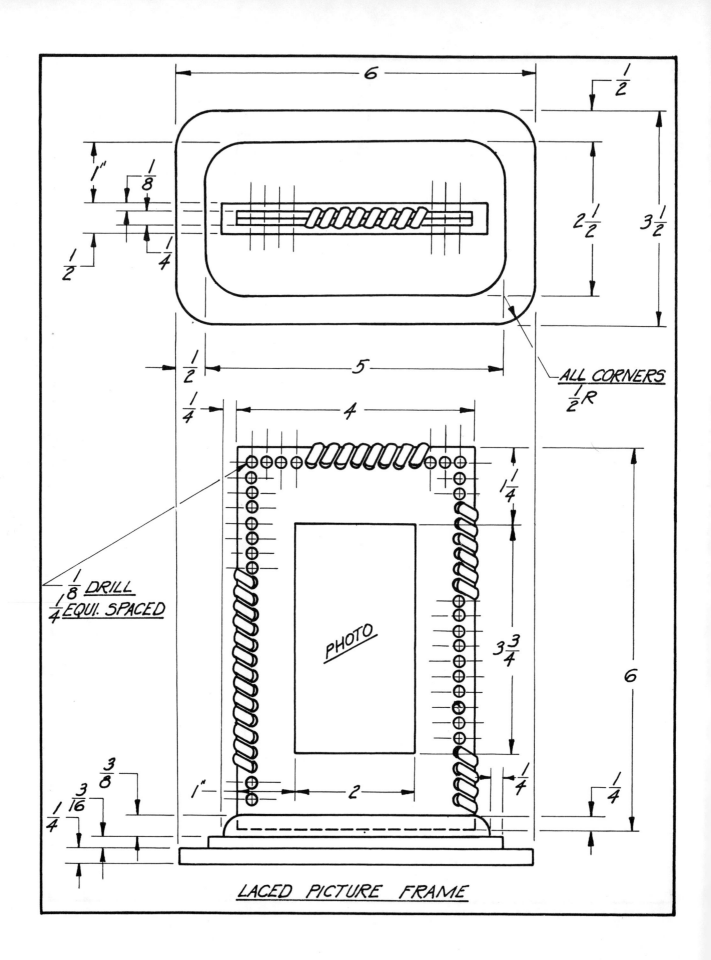

LACED PICTURE FRAME

PICTURE FRAME

SUBMITTED BY:
Harry Ruff
Instructor, Industrial Arts
Coffeyville, Kansas

MATERIAL REQUIRED:
1 piece white opaque plastic 1/4 x 3-1/2 x 6 in.
1 piece blue opaque plastic 3/16 x 2-1/2 x 5 in.
1 piece white opaque plastic 3/8 x 1/2 x 4-1/2 in.
2 pieces white opaque plastic 1/8 x 4 x 6 in.
Approx. 3 yards plastic lacing, blue opaque, 3/32 in. wide

PROCEDURE:
1. Cut all 4 flat pieces of plastic to size.
2. Radius all four corners of two base plates to 1/2 in.
3. Cut channel mount strip to size (item #3), and using table saw, saw channel into 1/2 in. wide face, 1/4 in. wide and 1/4 in. deep. If picture to be used is rather thick, cut channel slot slightly wider accordingly. Round off ends to pleasing contour.
4. Tape the two pieces of 1/8 in. white plastic 4 in. x 6 in. together using Scotch tape over edge on all four sides, or better yet, by using double sensitive masking tape between two pieces.
5. Lay out hole drilling placement along edge. Holes are placed 1/4 in. in from edge and 1/4 in. apart. Lay out center cutout section 2 in. x 3-3/4 in. (or any suitable size cutout to fit your picture).
6. Drill hole for entrance of scroll saw blade, and saw out center portion slowly and accurately. When this is completed use fine mill file to file edge straight, and smooth.
7. Drill 1/8 in. diameter holes around three edges as laid out previously.
8. Polish edges of all panels on buffer. Note: Inside edges of picture cutout may be solvent polished if desired. Also, inner

groove portion on channel need not be polished.
9. Remove all masking papers.
10. Seal two base portions together, then seal center channel to secondary base centered.
11. Place picture between two 1/8 in. panels, and lace edges with plastic lacing. Start at bottom corner and tie behind at opposite bottom corner.
12. Clean and wax.

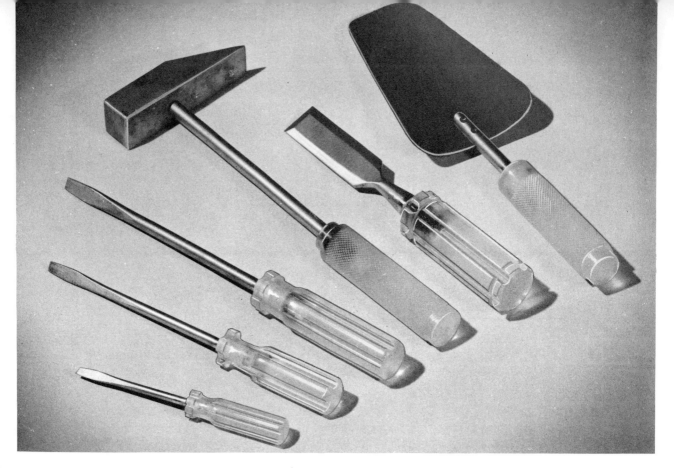

TENITE TOOL HANDLES

SUBMITTED BY:
Edward DeMello
Sequoia Union High
Redwood City, California

In a preceding chapter we mentioned one of the "cellulosics" made by treating cellulose (from cotton linters or wood pulp) with acetic and butyric acids to form the thermoplastic material called cellulose acetate-butyrate (or CAB). One trade name for this material is Tenite Butyrate (Tenite is a registered trade name of the Tennessee Eastman Co.) or sometimes "Tenite II." This plastic-like material makes excellent handles for tools because it is not brittle and may be struck repeated blows with a hammer without breaking. It is attractive in appearance, the natural color being a light, transparent or translucent yellow. However, the material may be dyed in manufacture to afford a number of different colors. Tenite softens at comparatively low temperatures, and is subject to attack by a number of common solvents, but neither of these factors is of importance to its use as tool handles.

Handles for steel shaft hammers, for screwdrivers, and other tools with metal shafts are easy to attach and will stand up under hard usage. Because the material is thermoplastic to a marked degree, attaching the handle is comparatively easy, and the results are excellent.

MATERIAL REQUIRED:
1 piece of Tenite rod, of suitable length and diameter for the tool to which it is to be applied. This rod is obtainable in round or fluted shaped extrusions. The round rod is usually used for hammers and chisels--the fluted for screwdriver handles and other applications where it is necessary to get a "turning" grip on the tool handle.
1 tool (to which the handle is to be applied) having a metal shank onto which the handle may be fitted.

PROCEDURE:
1. Prepare the metal shank of the tool

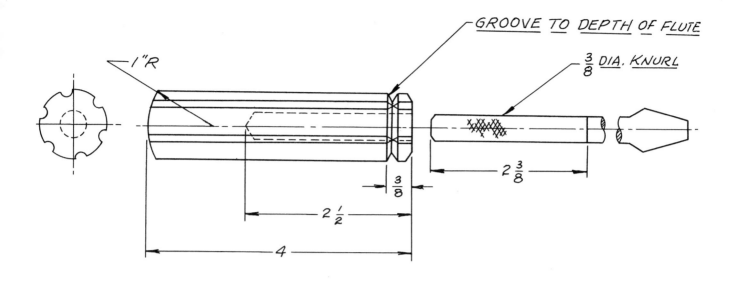

TENITE TOOL HANDLE

handle for inserting into the Tenite piece. Best results are obtained by knurling the part of the metal shank that is to be in the handle. This is to prevent slipping of the shank in the hole when the handle is twisted, and also to prevent the shank from pulling out of the hole. Of course, if the shank is square, hexagonal or shape other than round or smooth, this does not apply.

2. Prepare the handle "finish." If round bar-stock is used, it may be well to roughen the handle by mounting it in a lathe and using a regular knurling tool. One may turn down a step-off on one or both ends of the handle. In the case of a screwdriver, it is usually best to round off the end of the handle slightly. This may be done on a lathe, with hand tools, or abrasive paper. Tenite is soft enough to be cut rather easily with a good, sharp knife--this will enable the experimenter to "whittle" off the corners or shape a step-off, if he wishes. More even, and better looking work can be done on a lathe.

3. Drill hole in piece of Tenite stock to receive shank. If the shank has been knurled, drill the hole the same diameter as the shank. If a smooth shank is to be used, drill the hole in the plastic handle 1/64 in. smaller than the shank. For a square or hexagonal shank, drill the hole the same diameter as the distance measured across opposite flat surfaces, disregarding the protruding corners.

4. As soon as the shank and the handle are ready, the two may be assembled. First heat the shank with a blowtorch or similar flame. Get it hot enough so it will quickly melt a small piece of Tenite held against it. As soon as the shank is hot enough, press-fit it into the hole already drilled into the handle. The easiest way to do this in the shop is to mount the metal shank in the drill chuck of the drill press. Then set the handle, hole up, on the table of the press and bring the shank squarely down into the hole, exerting enough pressure to force it into the hole as far as necessary. The hot metal will melt the Tenite enough to allow it to be forced into the hole without too much pressure, providing the approximate dimensions given above for hole and shank have been observed. Allow to remain in the press for several minutes, until the plastic has reset due to cooling.

5. Remove from the press. Cut or file off any Tenite that has been forced out of the hole by the hot shank.

6. To give the Tenite a final, finished effect, the tool handle may be waxed, then lightly buffed.

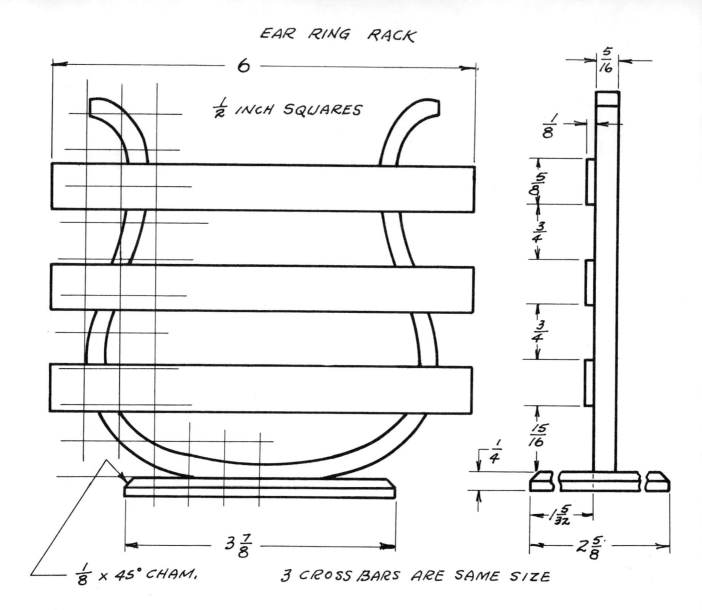

HARP-SHAPED EARRING RACK

MATERIAL REQUIRED:
 1 piece clear plastic 1/4 x 2-5/8 x 3-7/8 in.
 3 pieces colored plastic 1/8 x 5/8 x 6 in.
 1 piece clear plastic 1/4 x 5/16 x 14 in. long

PROCEDURE:
1. Cut and bevel piece of clear plastic 1/4 in. x 2-5/8 in. x 3-7/8 in. Beveling should be at 45 degree approximately one-half the thickness of the material.
2. Cut three strips of colored plastic 1/8 in. x 5/8 in. x 6 in.
3. Cut one strip of material 1/4 in. thick clear plastic x 5/16 in. wide x 14 in. long.
4. Sand and polish all five pieces.
5. Remove masking paper from all pieces.
6. Place the 14 in. long strip of material in oven for about four minutes at approximately 275 degrees and when flexible, remove, and free form into harp-shaped design such as drawing indicates. A forming jig may be used for making identical pieces over and over again by the use of some 1/4 in. thick plywood in this case. Fabrication of the forming jig makes an ideal woodworking project.
7. After strip is formed into harp-shape, touch bottom of shape to disc sander set

Projects From Plastics

at 90 degrees in order to flatten off an area about 1 in. wide for flat sealing to base panel.

8. Using small camel hair brush, spot some solvent at points where the colored strips will touch harp-shaped strip. After keeping spot wet for at least thirty seconds, place plastic strip in position. This should be done one strip at a time (spotting and all).

9. After the three colored strips are in position and are thoroughly dry, using soak method, place flattened area of harp-shape in Pleximent, allow about forty-five seconds for soaking, then place in position on base for sealing. Allow to dry thoroughly.

10. Clean and wax.

Harp-shaped earring rack.

NAPKIN HOLDER

This unique napkin holder holds a regular size package of paper napkins in place, and allows them to be removed with ease.

MATERIAL REQUIRED:
1 piece 1/4 x 7 x 8 1/2 in. clear acrylic for base.
1 piece 3/8 x 1 x 8 1/2 in. clear acrylic for hold-down bar.
2 pieces 1/2 in. dia. clear acrylic rod 5 in. long.
4 only, clear acrylic molded "domes."

PROCEDURE:
1. Cut all parts to exact size.
2. Polish all edges of sheet material and both ends of rods.
3. Drill (2) 1/2 in. dia. holes on 7 1/4 in. centers in base plate to accommodate the 1/2 in. dia. rods.
4. Using plastic solvent, cement the two rods into base.
5. Drill (2) 9/16 in. dia. holes in acrylic hold-down bar on 7 1/4 in. centers.
6. Countersink holes from both sides to approximate 1/8 in. depth.

This oversize hole allows the hold-down bar to slip over 1/2 in. dia. rods in base with ease, thus allowing ease of removal. Hold-down bar may be beveled on all edges before polishing if desired.

Using plastic solvent, cement 4 small molded acrylic domes in corners as shown.

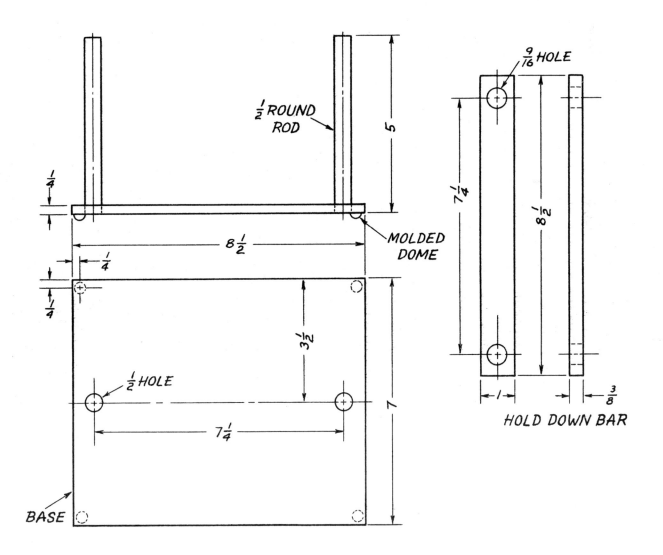

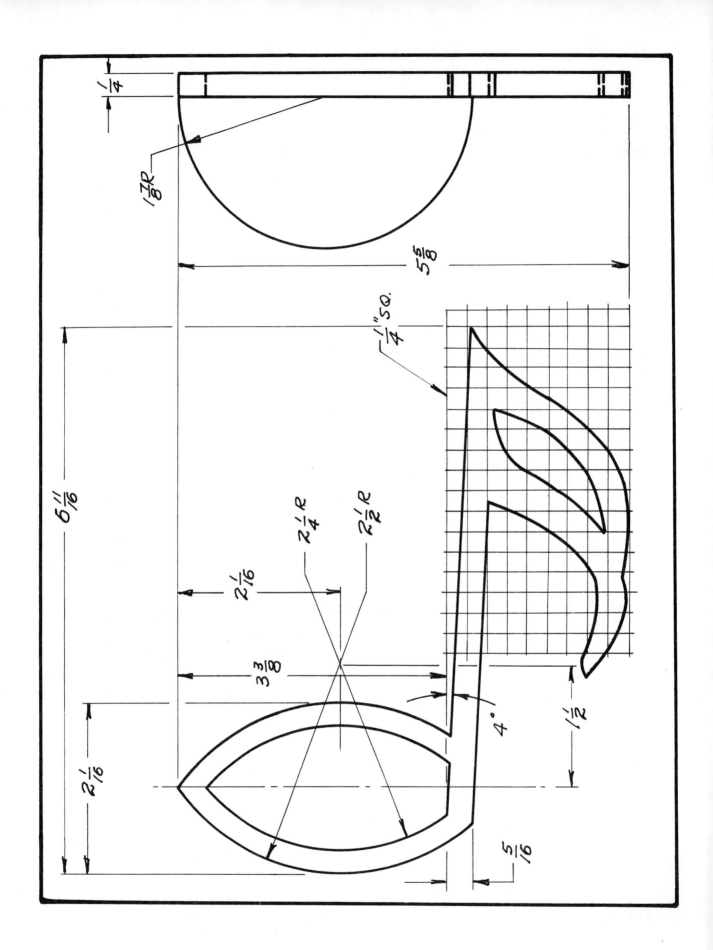

MUSICAL CLEF DECORATIVE SHELF

SUBMITTED BY:
V. S. Fox
Lincoln High School
Cleveland, Ohio

This little "what-not" shelf, made of clear plastic, is a decorative item, which is particularly appropriate for the music room, or den of a music lover. The curved lines should be cut very accurately.

MATERIAL REQUIRED:
1 piece clear plastic 1/4 x 5-7/8 x 7-7/8 in.
NOTE: If properly laid out, both parts of this project may be cut from a sheet of the given dimensions. Size of shelf may be varied as desired, provided all dimensions are kept in the proportions shown in drawing.

PROCEDURE:
1. Lay out design on paper and make patterns, or use carbon paper and pencil to trace design onto paper covering of plastic.
2. In order to insert blade of jig saw, drill two holes about 3/8 in. in diameter in waste stock.
3. Center cutouts should be done before cutting outer outlines to allow more area for holding while sawing.
4. Saw outer outline cuts of clef. Saw out shelf piece from unused portion of original sheet. For best symmetry of finished design, shelf should be of semi-circle shape.
5. Sand all edges, external and internal. In this particular piece, edges of stem, and outer edges of flag and note body should be rounded off considerably. Body of note should be elliptical in over-all shape.
6. Buff and polish all sanded edges of clef that are accessible. Solvent-polish inaccessible edges by applying solvent sparingly with small camel hair brush.
7. Sand and polish all edges of small shelf.
8. Cut surface designs in shelf and clef if desired.
9. Bond shelf across oval length of clef body, as shown, using EDC solvent for cementing.
10. Clean and wax.
NOTE: Shelf should be affixed to wall using screws for hangers.

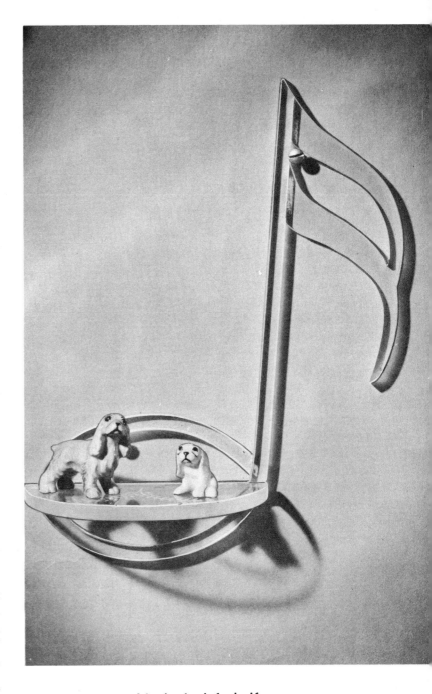

Musical clef shelf.

PIPE HOLDER

SUBMITTED BY:
Darwin Eaton
Industrial Arts Department
Swanton, Ohio

MATERIAL REQUIRED:
1 piece clear plastic (or colored) 1/4 x 3 x 7-3/4 in.
2 pieces clear plastic (or colored) 1/4 x 1-1/2 x 5 in.
4 pieces clear plastic (for legs) 1/2 x 1/2 x 1/4 in.
1 piece clear plastic 2 x 2 x 2 in.

PROCEDURE:
1. Cut 1 piece of plastic 1/4 in. x 3 in. x 7-3/4 in.
2. Cut 4 pieces of clear plastic 1/4 in. x 1/2 in. x 1/2 in. to serve as "legs" on above stand. You may wish to use #61 half-round domes instead of the squared off pieces of material, and if so, these can be purchased ready-made.
3. Cut 2 pieces for curved sections 1/4 in. x 1-1/2 in. x 5 in.
4. If the center decorative cube is to be added, which of course can be any carving, or other decoration, then cut a 2 in. cube of clear plastic.
5. Sand and polish the edges of all pieces.
6. Internally carve the cube to desired design, or decorate block as desired.
7. Lay out dimensions and locations on the two strips 1-1/2 in. x 5 in. x 1/4 in. for 3/4 in. diameter holes to be drilled through, and also for 3/4 in. wide slot to be cut into one end.
8. Drill 3/4 in. hole on drill press, making **sure** to back up the piece with a piece of scrap plastic or with wood backing, so as to not break out back side of hole when drill goes through the stock.
9. Cut 3/4 in. wide slot in ends of the two pieces, using band or scroll saw. Smooth inside cuts of slot with fine mill file, and sand smooth.
10. Form these two 1-1/2 in. x 5 in. strips to desired angle or arc of circle as described by heating in oven for approximately four minutes at 275 degrees F. (Be sure to remove masking paper first).
11. Remove masking paper from all other pieces.
12. Sand a "flat" on both pieces to be used for pipe supports at desired location. This will be approximately at spot where 3/4 in. hole comes through the panel, thus giving proper tilt.
13. Using Pleximent and soak method, seal the four legs to bottom of base piece.
14. Using soak method, and Pleximent, seal the two pipe supports to base.

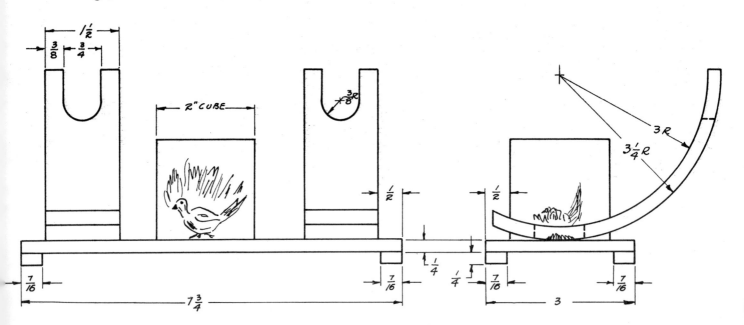

PIPE HOLDER

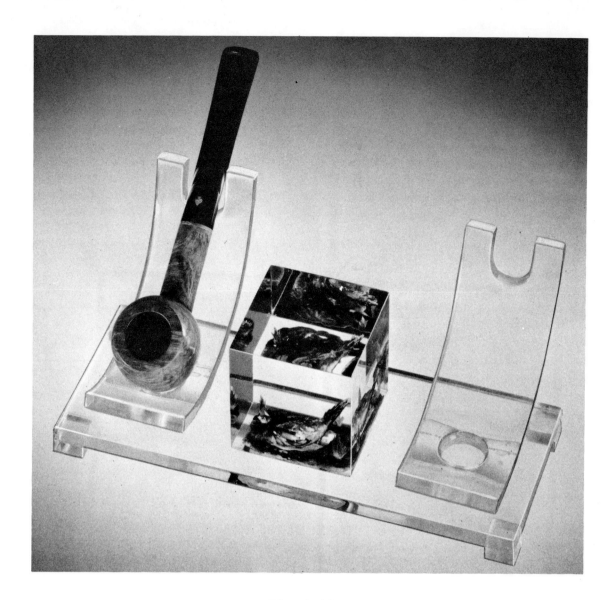

Pipe holder.

15. Seal cube or decorative block in place using Pleximent, soak method. (Be sure that carving has been filled with white filling plaster and dried hard, and also that it has been carefully smoothed on bottom). Clean and wax.

WHEELBARROW CANDY, NUT TRAY

SUBMITTED BY:
Clyde Brown
Carthage High School
Carthage, Illinois

This colorful tray, made up in the form of a wheelbarrow, is a suitable ornament for coffee or end table. The body of the wheelbarrow is cupped red transparent plastic. This is cemented to two side rails, with legs, and a wheel, all made of white, translucent plastic. The ornamental tray thus formed may be used for holding nuts, candy or other knick-knacks.

MATERIAL REQUIRED:
1 piece 1/4 in. white translucent plastic, to make circle 2-1/8 in. in diameter
2 strips, same material as above 3/16 x 1/4 x 7 in. for side rails
2 strips, same as above, 3/16 x 1/4 x 3-1/2 in. for curved legs
1 square, red transparent plastic, 1/8 x 4-1/8 x 4-1/8 in.
1 piece, 14-gauge bronze wire, 1 in. long, for axle

NOTE: Strips for side rails and legs may be cut from 1/4 in. stock by cutting three strips that will finish to 3/16 in. thick and 7 in. long.

PROCEDURE:
1. Saw wheel circle, and side and leg strips from white translucent plastic, 1/4 in. thick, cutting strips 3/16 in. wide.
2. Saw out square of red, transparent plastic, from 1/8 in. stock. Square is 4-1/8 in. x 4-1/8 in.
3. Place red square in oven at 275 degrees F. and also the two 3-1/2 in. long leg

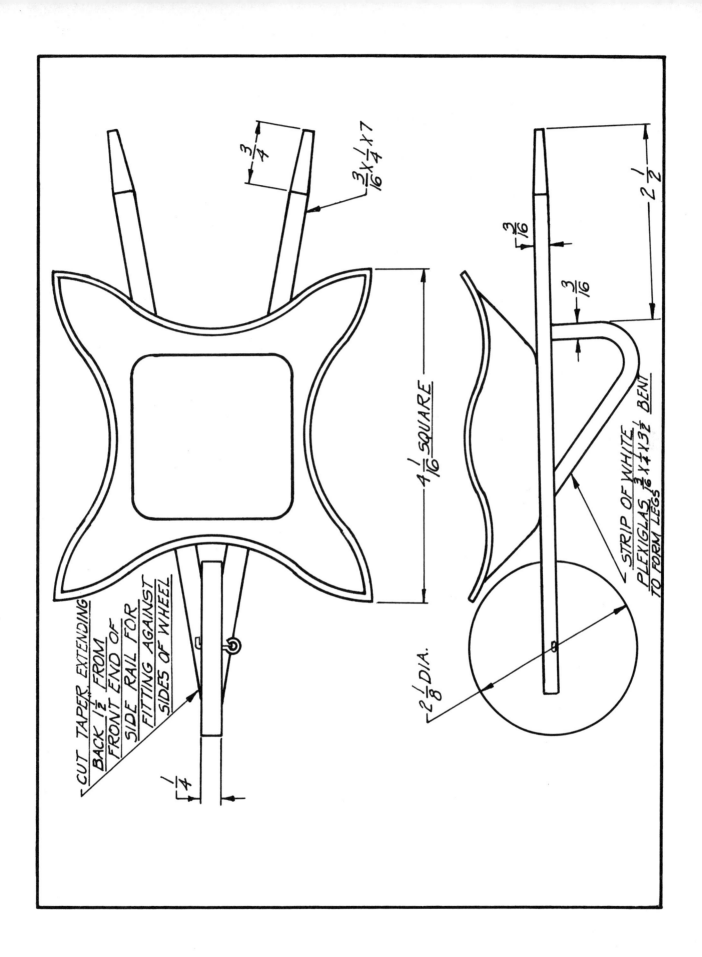

strips made from the white stock.

4. Cut long bevel on front end of each side rail strip. This bevel allows the side rail to fit parallel to sides of wheel while spreading out toward rear. The bevel, measuring on the quarter-inch face, begins at one corner of the end and comes back for about 1-1/2 in.

5. Round off and taper back end of side rail piece to form handle. This should begin about 3/4 in. from the end and may be done by scraping or sanding.

6. Remove the leg strips from the oven when soft enough to bend, and shape as shown in the drawing. Be sure bends in legs are identical.

7. Wet-sand the two ends of the bent leg strips on a sanding belt or on a flat surface so they will make flat contact with the side rail strips for cementing.

8. Cement the legs to the flat or 1/4 in. face of the side strips, making sure that the bevels of the side rails are on opposite sides--that is, so that both bevels will be adjacent to the wheel when the legs are in the down position.

9. When properly softened remove the red square from the oven and bend each edge up at the mid-point. Use a block of wood 1-1/4 in. square for the inside bottom of the bed piece, to be sure that it has a flat bottom surface in the center where it is to be cemented to the side-rails. The bends may be made by hand, but better results can be had if you use a jig made by fitting four dowel pins into a wood block so the dowels are at the corner of a square 1-3/4 in. on a side, measuring from the inside edge of the dowel pin, nearest the center of the square.

10. Sand, buff and polish all edges, except that part of the side rails where the bed piece will come in contact when cementing.

11. Drill small holes in the side rail pieces about 5/8 in. back from point of beveled end. Make hole equal to diameter of wire to be used as the axle. Drill hole of same size in center of wheel piece.

12. Fit axle through side rail holes and center hole of wheel. It would be best to use a small piece of fairly heavy paper as a spacer on each side of the wheel to allow enough space between wheel and side rail pieces so that wheel can turn.

13. Cement bed piece of red plastic to side rails. Place the bed so the center is about three inches back from pointed end of side rails, and so it is in the center of the V angle formed by the two diverging side rails.

14. Wax and polish the entire piece.

PIN-UP LAMP

SUBMITTED BY:
V. S. Fox
Lincoln High School
Cleveland, Ohio

This attractive and sturdy pin-up lamp is made of two colors of opaque plastic. The back or wall-piece is a keystone-shaped piece of white opaque plastic, the lamp bracket is green, and the brace strip for the bracket is white. Other color combinations are, of course, possible. The actual fabrication of the piece is relatively simple, and the result is well worth the time, material and effort involved.

MATERIAL REQUIRED:
1 piece opaque white plastic, 3/16 x 5-1/4 x 6 in. for back or wall plate.
1 strip of 3/16 in. white, 7/8 in. wide x about 10-1/2 in. long. The length of this piece is not critical, since the loops at each end can be made larger or smaller, and the brace may be set either in or out from the right angle of the lamp bracket, to place it in the proper position for its length.
1 strip of green opaque plastic, 3/16 x 1-3/8 x 10-1/2 in., for the lamp bracket piece
1 short piece green plastic strip, 1-1/8 x 1-3/8 in. and 3/16 in. thick. This piece is cemented to the top of the bracket to strengthen it at the point where the lamp socket is fastened.
1 lamp socket with attaching brass pipe nipple and nut, together with cord and plug for plugging into wall outlet.
1 Shade (not shown).

PROCEDURE:
1. Cut out back or wall plate, according to dimensions given in drawing. This plate is 6 in. top-to-bottom. The sides slant outward from bottom, where the plate is about 3 in. wide, to the widest point of 5-1/4 in. wide at a point about 5/8 in. down from the top line of the sheet. A slot 3/16 in. wide is cut up from the center point of the bottom edge, and perpendicular to the bottom line, for a distance of 4-1/4 in. This slot is to accommodate the fixture wire where it passes up and comes through the slot above the angle of the brace, to enter the socket. If the wire is fed down through this slot it allows the wall plate to rest flat against the wall. The top corners of the wall plate are cut at an angle, as shown, with the bevel cut making approximately equal angles with the top line and the sloping lines. Sand and polish all sawed edges of this plate, also edges of the green and white strips.
2. Remove the protective paper from both sides of the green and white strips, clean the surfaces, and heat in oven to 285 deg. F. until soft enough to bend. The points of bending may also be softened over a strip heater, if the piece is held far enough away from heating element to allow a sufficient length of the strip to become flexible. Or by careful handling the pieces can be heated at the proper points over an electric hot plate.
3. Bend a loop on each end of brace strip (the white strip) as shown in drawing, keeping the mid-portion of the piece flat and straight. These loops may be a freehand bend, though perhaps more uniform results are obtainable if they are bent around a 3/4 in. dowel rod. The loops will be more attractive if the last half-inch at the end of the strip remains straight, and meets the straight portion of the brace strip at right angles, as shown.
4. Make the "U" bend at the top of the green bracket strip according to the approximate dimensions shown on the drawing, being sure the short end of the "U" is parallel to the other leg of the bend.
5. Bend the wide right angle bend of the bracket strip. This may be done by lightly clamping the strip at the proper point across a 1-inch strip of wood, then bending the two legs of the bracket angle until they are at a right angle or 90 degrees to each other. This leaves the lower or straight leg of the bracket with about 3 in-

Sturdy pin-up lamp.

ches of flat area before the bend begins. This flat area is to be cemented to the back plate in a later step.
6. Cement the reinforcing short green piece to the top of the "U" bend.
7. Drill a 7/16 in. hole through this doubled portion of the bracket, centering the hole about one-half inch back from the end of the strip.
8. Cement the bottom leg of the bracket strip to the back plate. Center the strip exactly in the center of the plate, cementing it over the groove or slit that has already been cut in the back plate. The bottom of the green strip should be about 1/4 in. above bottom line of back plate.
9. Sand "flats" on the brace strip (white strip that has been looped at the ends) at the points whre this strip will be cemented to the lamp bracket. This brace should be cemented in such a way that it will be at an angle of approximately 45 degrees to the back plate. If everything has been done properly up to this point, the two flats sanded on the looped ends should be in planes that are perpendicular to each other, as the wall is to the floor in a room.
10. Cement the brace in place, making it as near to a 45 degree angle with the back as possible.
11. After all the cemented joints are dry, polish and wax the final job.
12. The lamp may be hung on a hook in the wall with the hook extending into the top of the slot in the back plate. For best results a shade should be put over the bulb in the fixture.

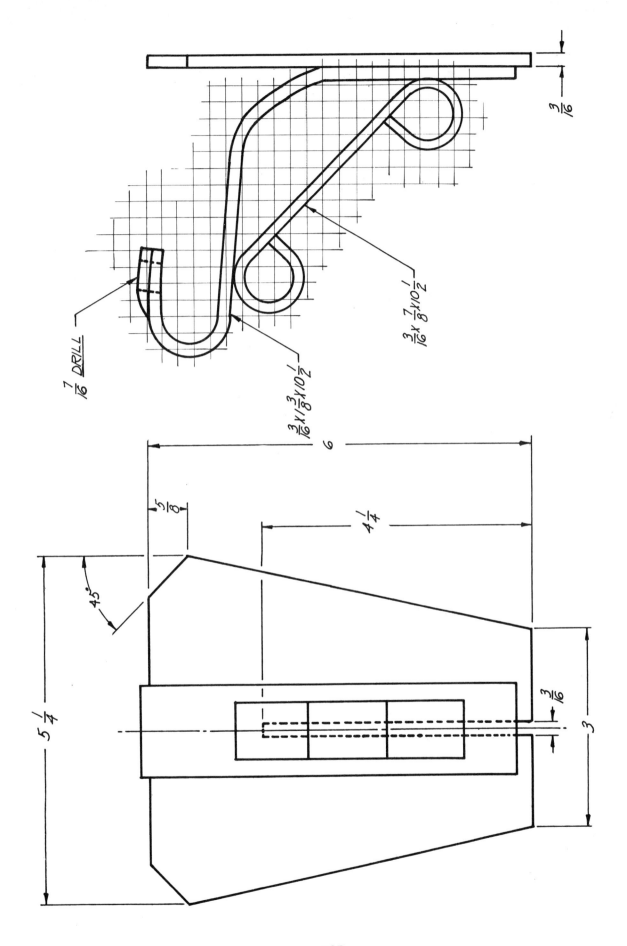

Clef bud vase.

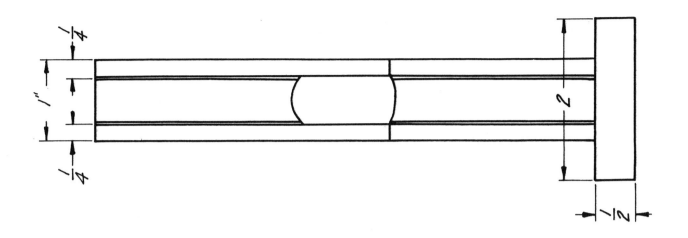

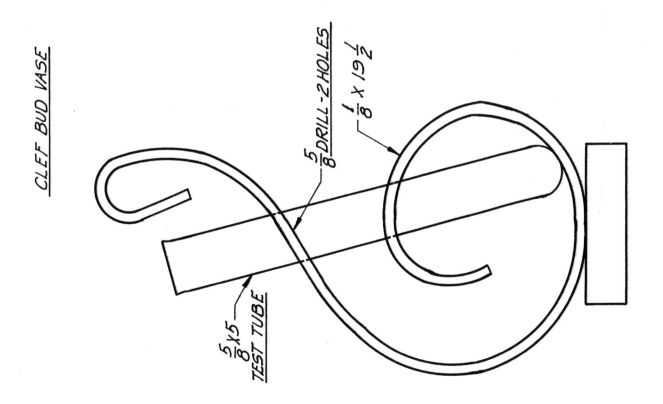

CLEF BUD VASE

SUBMITTED BY:
Phil Brooks
University of Missouri
Columbia, Missouri

Here is one of the most attractive little items that can be made from crystal clear plastic. The amount of material involved is small, yet, when complete, makes into a very colorful item. The piece of material which is formed into the clef shape can be made of clear or colored plastic, thus, color is optional. A very attractive idea is to use one of the fluorescent colors such as green fluorescent for the strip. When the decorative saw cut is made, the resultant cut line has the brilliant fluorescent appearance.

MATERIAL REQUIRED:
1 piece clear plastic 1/8 x 1 x 19-1/2 in.
1 piece clear plastic 1/2 x 2 x 2 in.
1 glass test tube 5/8 in. diameter x approx. 5 in. long

PROCEDURE:
1. A forming jig is necessary for this project in order to produce matching sets or more than one of these holders. This jig can be made of 3/4 in. plywood, using a base of approximately 5 in. x 8 in. Onto this, nail the blocks of 3/4 in. plywood cut to shape to form the outline of the clef. Allow spacing of approximately 5/32 in. so that 1/8 in. material will slip into position easily. This jig is also necessary for the drilling of the 5/8 in. hole later.
2. Cut the strips of plastic to size, 1/2 in. x 2 in. x 2 in., and the strip to size, 1/8 in. x 1 in. x 19-1/2 in.
3. Sand and polish all edges of both pieces.
4. Set guide on circular saw so that a cut can be made approximately 1/4 in. from edge on the long strip, and raise the blade to about 1/32 in. above saw table. Then saw the two decorative lines into the strip of plastic, one from each long edge.
5. Remove all masking paper and clean parts.
6. Place strip in oven set at about 275 deg. F., and heat for about 3-5 minutes. When soft and pliable, remove, and place in wood jig starting with top end first. Bend into position as indicated and allow to cool.
7. With the part still in forming jig, locate position of hole to take test tube. Drill through wood (edge), and plastic. Drill hole 5/8 in. diameter, or to match size of test tube used.
8. Remove from jig, and sand slight flat on bottom of clef to facilitate bonding to base.
9. Using soak method, seal clef to center of base and allow to dry.
10. Clean and wax.
11. Insert test tube for flowers.

DESK LIGHTERS

SUBMITTED BY:
A. LeRoy Kahler
Bloomsburg, Pennsylvania

The lighters in this project picture are mounted in two types of bases--one is laminated colored plastic--the other is an internally carved block with colored backing, or base. Colorful combinations of plastic can be laminated to give a striking appearance. (Black and white plastics were used in making lighter shown in photo). In most cases one can make a striking combination by using white opaque plastic with any other opaque dark color--red, black, blue, green, etc. Of course, internal carving always is in proper taste when colors are properly displayed with the carving. The barrel-shaped laminated block in this project is just an idea for a unique design and can easily be varied as to shape and size.

MATERIAL REQUIRED:
(a) For Barrel design:
6 pieces white opaque plastic 1/4 x 2-1/4 x 2-1/4 in.
6 pieces black plastic 1/4 x 2-1/4 x 2-1/4 in.
1 Evans lighter (short model) (1-1/2 in. diameter base)
(b) For Carved design:
1 piece clear plastic 2 x 5 x 5 in.
1 piece black plastic 1/8 x 5 x 5 in.
1 Evans lighter (short 1-1/2 in. model)

PROCEDURE:
(a) For Barrel design:
1. Saw the 12 pieces of colored plastic to size. The 2-1/4 in. pieces should be this size to begin with, as they will be laminated and turned to a smaller diameter. Thus, this 2-1/4 in. square allows adequate size for cleaning up to 2 in.
2. Remove masking paper from all pieces.

Desk type cigarette lighters.

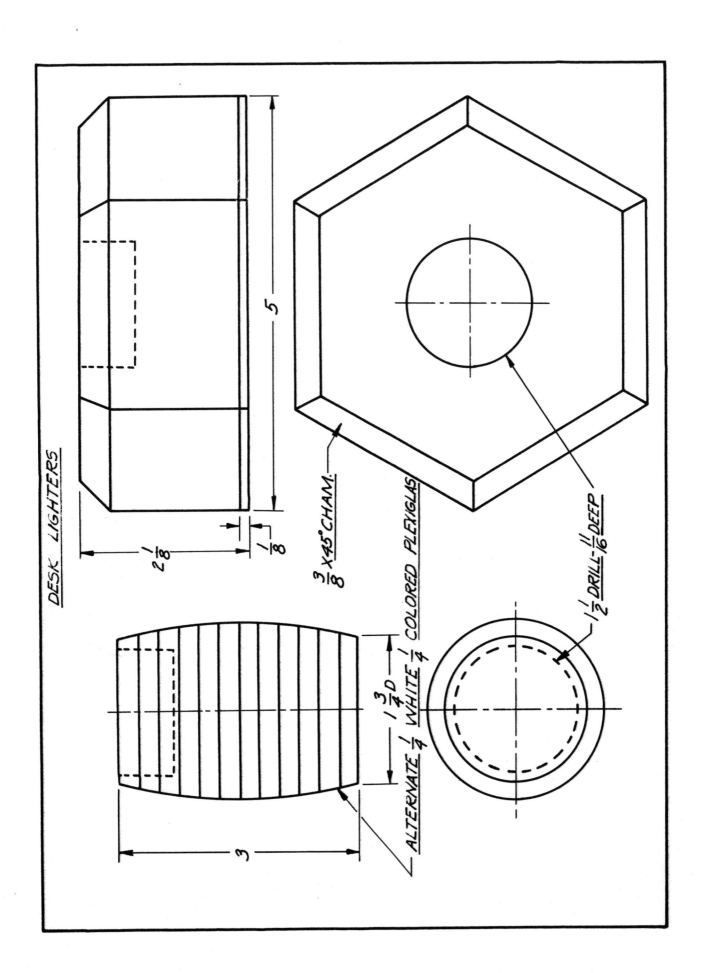

3. Using soak method, seal all twelve pieces together, alternating colors as indicated in photo. Allow to thoroughly dry.
4. Either sand to barrel shape, or if lathe is available, chuck the block in lathe and then turn to barrel shape. Ends should finish at about 1-3/4 in. diameter, and largest center point at about 2 in. or slightly over. Lathe turn in four jaw chuck as close as possible, then reverse ends, and use three jaw chuck to hold block for finishing of opposite end. While in lathe, sand smooth with #320-400 wet-or-dry paper.
5. Chuck block in lathe and drill 1-1/2 in. diameter hole 11/16 in. deep to receive lighter unit.
6. Polish outer surface of block with buffer.
7. Clean and wax.
8. Insert lighter into block.

(b) For Octagon internally carved block:
1. Cut 2 in. thick block to size, then cut corners as indicated.
2. Cut 1/8 in. black base to size, then cut corners as indicated.
3. Remove masking paper from both pieces.
4. Internally carve, dye, and fill block as desired.
5. Make sure internally carved surface is absolutely flat, then soak and seal to black base. Allow to dry.
6. Lay out center of top of clear block.
7. Drill 1-1/2 in. hole 11/16 in. deep to receive lighter.
8. Bevel top of 2 in. block, and sand all edges on disc sander to even size and octagon shape.
9. Wet-sand and polish all edges (sides).
10. Clean and wax.
11. Insert lighter unit into base.

NOTE: The hole to receive lighter may be solvent polished if desired by pouring ethylene dichloride (Pleximent) into hole, allowing it to remain for about fifteen seconds, then pouring out, and allowing it to dry . . . without touching it. This should be done before final buffing operations, as buffing will be needed to remove any marred surface caused by solvent. Otherwise, hole may be left in frosty drilled condition which is not objectionable if properly done.

CARVED-LID JEWEL BOX

SUBMITTED BY:
George Feuerstein
Winsted, Connecticut

This small jewel box of clear plastic, fitted with a hinged, internally-carved lid, is one of the most beautiful objects included in this series of projects. The carved roses and leaves, appropriately dyed, in the lid are set off by giving the lid a backing of black enamel followed by a covering of wool flock. The effect of the three-dimensional flowers and leaves standing up in the plastic, against the opaque, black background is startling and beautiful.

MATERIAL REQUIRED:
1 piece 1/8 in. clear plastic 3-1/4 x 4-3/4 in.
2 pieces 1/4 in. clear plastic 3-7/8 x 1-5/8 in. for box ends
1 piece 1/4 in. clear plastic 2-13/16 x 3-1/4 in., box lid
2 brass pins, 1/16 in. dia. x 3/8 in. long, for hinge pins

PROCEDURE:
1. The front, back and bottom of the box are made by bending the larger piece of 1/8 in. plastic, as shown in drawing.
2. After bending, the two ends of the bent piece must be wet-sanded on belt sander or flat surface. These surfaces are to be bonded to the two end pieces, as shown, and must make contact at all points.
3. The box lid is carved by procedures described in the section on Internal Carving. The rose and leaf carvings are dyed and filled in from the back, and the back or bottom side of the lid is then given a coat of black enamel. Wool flock is blown onto the wet enamel for additional opaqueness.
4. Hinge pin holes are located in the ends of the lid and the end pieces of the box. Drilling must be done carefully so the lid will fit down within the box ends and the top of the lid will be flush with the top edges of the ends.
5. Hinge pins are inserted in the holes and the two end pieces are cemented to the ends of the curved 1/8 in. piece. Make sure the front of the box body exactly meets the lid, when lid is closed. Also, be certain the top edge of the back of the box does not come up high enough to prevent opening of lid. As shown in drawings, the back lower edge of lid is beveled off to allow for easier opening of the lid. The box should be assembled with the two end pieces standing on edge on a flat surface and with the bottom of the curved body of the box also resting on this flat surface.
6. After the cement is dry, wax and polish the entire box.

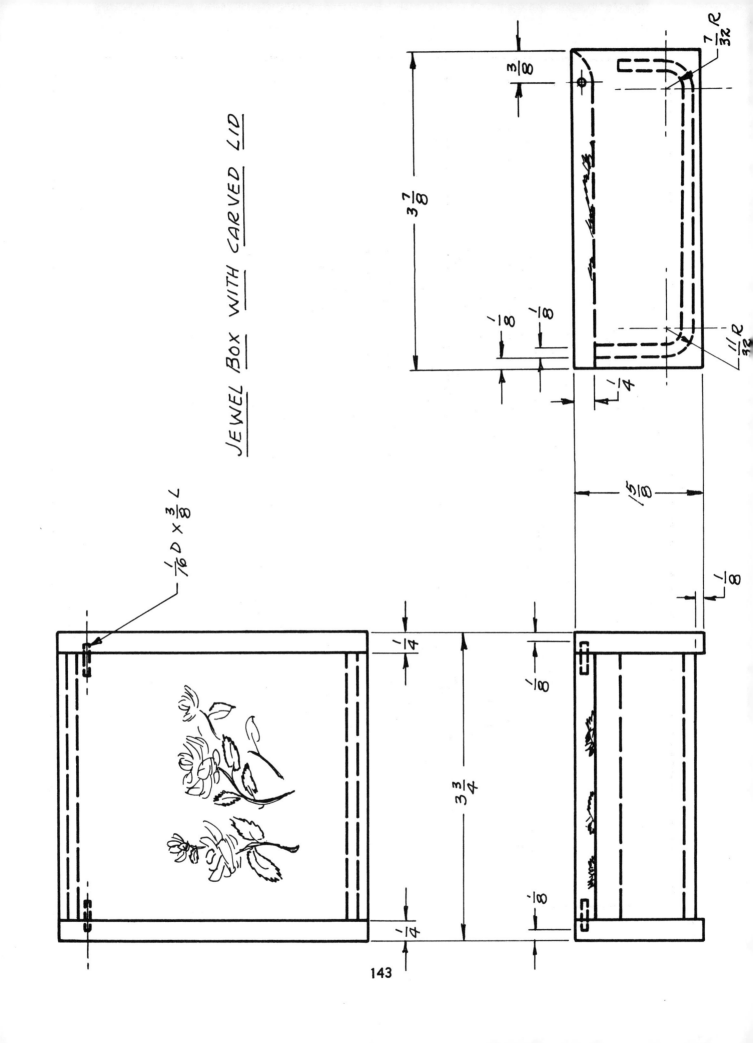

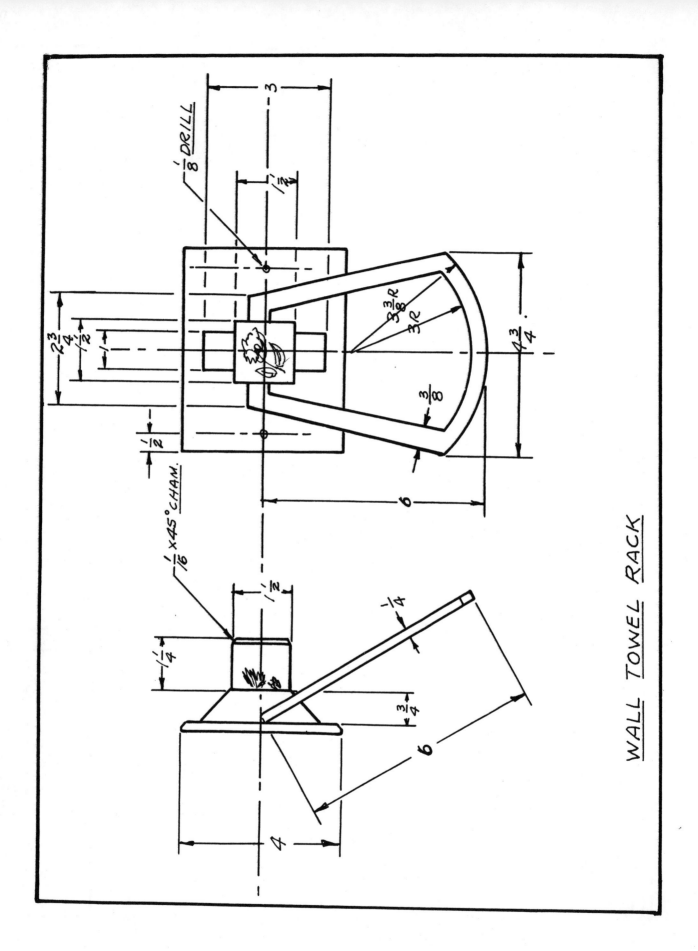

WALL TOWEL RACK

SUBMITTED BY:
A. LeRoy Kahler
Bloomsburg, Pennsylvania

When fully polished to its gleaming beauty, this towel rack will make a welcome addition to any bathroom or kitchen.

MATERIAL REQUIRED:
1 piece 1/4 x 4 x 5 in. white opaque plastic
1 piece 3/4 x 1 x 3 in. black plastic
1 piece 1/4 x 4-3/4 x 6 in. black plastic
1 piece 1-1/2 in. diameter cast clear plastic rod 1-1/2 in. long

PROCEDURE:
1. Cut white back plate to rectangle shape and bevel edges at 45 degrees.
2. Drill 1/8 in. holes at points indicated for screw mount.
3. Cut 3/4 in. thick black block to shape, and make half round cutout in back flat side as indicated.
4. Cut 1-1/2 in. diameter rod to length, sand and bevel, and sand 1 in. wide flat lengthwise as pictured in top view.
5. Cut swinging bracket to shape on band saw, and, using file, round off two ends as indicated, thus allowing free swinging in black half round cutout.
6. Sand and polish all edges, and also ends of rod.
7. Remove all masking papers.
8. Internally carve rod block, dye, fill, and allow to dry.
9. Seal black block to indicated position, and after dry, seal carved rod block into position and allow to dry.
10. Twisting ends apart slightly, install hanging bracket into black back plate.
11. Clean and wax.

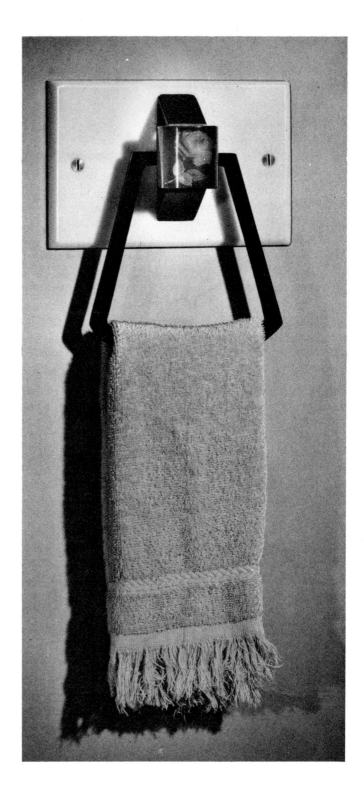

COMPRESSED CUBE CARVING

SUBMITTED BY:
Phil Brooks
University of Missouri
Columbia, Missouri

Compressed cube carving is a process which provides unusual results. "Curved" lines and "curved" carvings can be achieved in blocks of Plexiglas, which dramatically illustrate the "plastic memory" principle, or its ability to return to its original shape after being heated and formed, then reheated. An unusual appearance that one has carved "curved" lines into the block will result, which generally creates an air of mystery. Following is the procedure for accomplishing this unusual result.

MATERIAL REQUIRED:
1 only clear Plexiglas cube 2 x 2 x 2 in.

PROCEDURE:
1. Remove all masking paper from the Plexiglas cube, and clean.
2. Sand and polish to crystal clear state.
3. Place the block in an oven at approximately 280-300 degrees F. for approximately 45 minutes. Since the material itself is a good insulator, it requires even, prolonged heating until the block is heated through to its center core. The block will have to be handled with a pair of soft cotton gloves, as it becomes too hot to handle otherwise.
4. When thoroughly heated, remove from oven and place between two pieces of of wood (hardwood with smooth surface) and compress in vise until as flat as possible (center photo). Allow to cool in this position in vise.
5. When thoroughly cooled remove from vise and drill four holes as shown in drawing--making sure not to drill completely through opposite surface. A little experimenting here with hole placement can result in some very startling final appearances.
6. After holes are drilled, place the compressed block back in the oven at same temperature, and allow sufficient time (usually about 30 minutes) for the block to return to its original cube shape.
7. If any surface marks from clamping have resulted, then the cube should be resanded and rebuffed.
8. At this point, the regular internal carving can be done in the block in the center area between the curved lines, then all curved lines and carving can be dyed and filled. The entire process is an interesting one, and will create a novel effect.
9. Clean and wax.

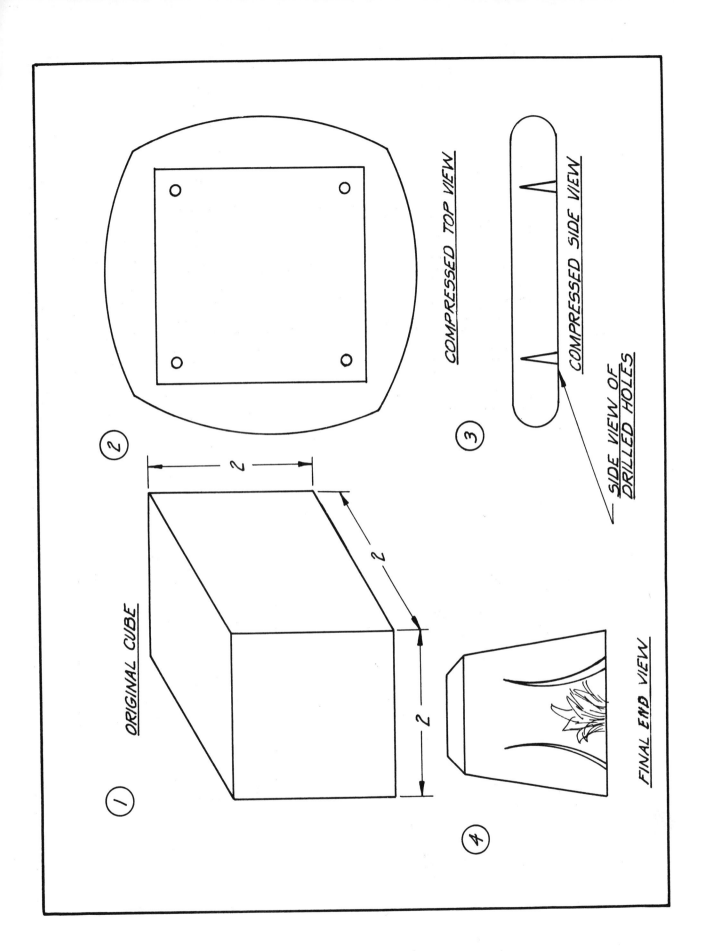

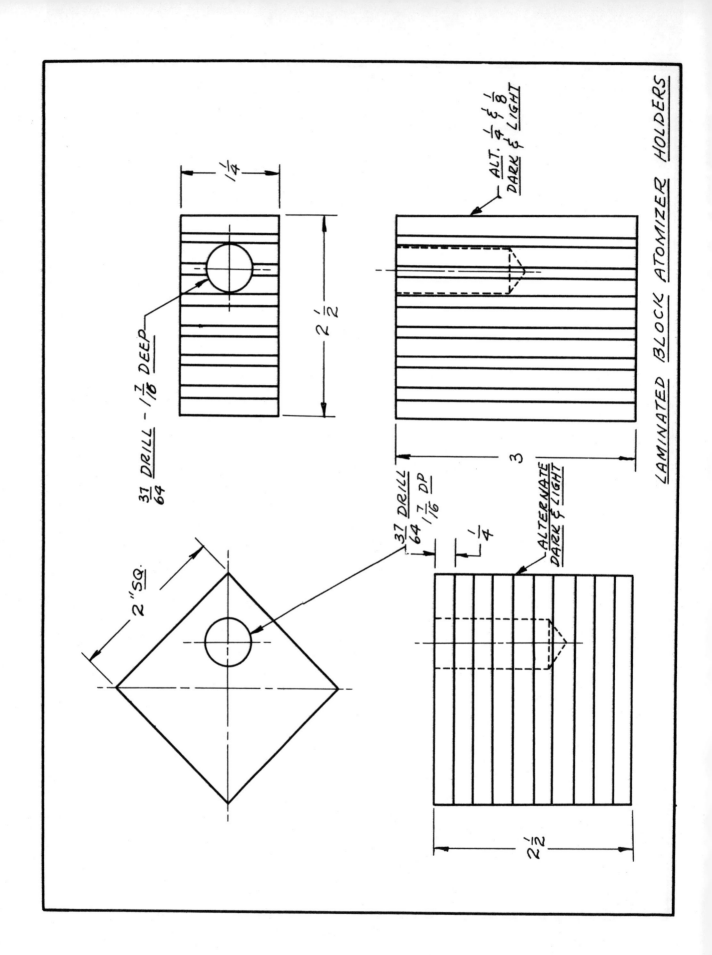

LAMINATED BLOCK PERFUME BOTTLE HOLDERS

SUBMITTED BY:
A. LeRoy Kahler
Bloomsburg, Pennsylvania

Colorful blocks built up by laminating together colored plates of plastic make highly attractive perfume bottle or atomizer holders. The construction of these blocks is quite simple and the results are well worth the small amount of effort required to make them. After laminating the plates together, a hole of suitable size is drilled into the laminated block to accommodate the perfume atomizer bottle.

MATERIAL REQUIRED:
Square, rectangular or other shaped pieces of the desired colors. The blocks should be of identical shape, or nearly so, and of proper thickness to achieve the desired effect.

PROCEDURE:
1. Prepare the pieces that are to go into the finished laminate so that they are of the same size.
2. Cement these pieces together by any suitable cementing process, so that the corresponding edges coincide as nearly as possible. This is done to cut down the amount of finishing needed on the laminated block. Clamp the cemented parts lightly together and follow procedure previously described for cementing the plastics together.
3. After the cement has dried, wet-sand all faces of the figure until perfectly flat and smooth--then buff and polish.
4. Drill hole of suitable size and depth in the block to accommodate the glass vial of the atomizer.
5. Wax and polish.

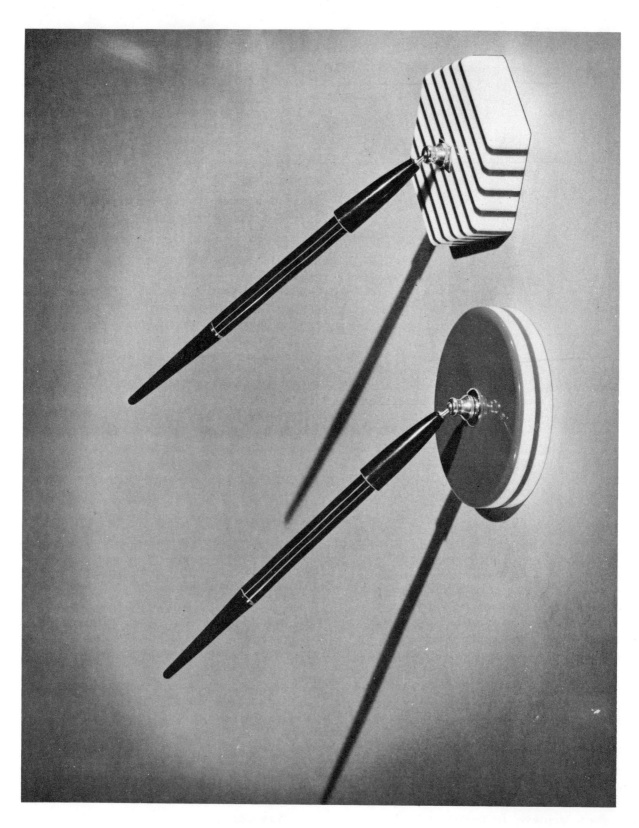

Laminated pen sets.

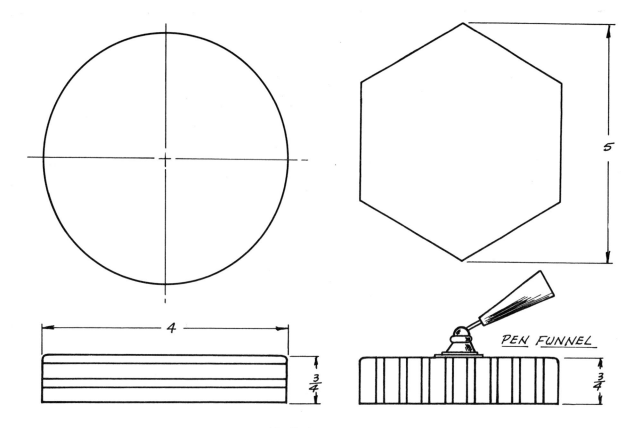

LAMINATED PEN SETS

SUBMITTED BY:
A. LeRoy Kahler
Bloomsburg, Pennsylvania

A desk set with an attractive laminated base makes an ideal gift for the office worker.

MATERIAL REQUIRED:
(For fabricating hexagonal base pictured on right)
13 pieces black opaque (or any suitable color) plastic 1/8 x 3/4 x 5 in.
14 pieces white opaque plastic 1/4 x 5 x 3/4 in.
1 Fountain Pen and socket-swivel
1 threaded stud (or bolt) to fit thread on swivel

PROCEDURE:
1. Cut 27 strips of plastic to size per required list.
2. Remove masking paper from all strips.
3. Using soak method, and Pleximent, laminate alternate colors of the material, starting on the outside with a strip of the 1/4 in. thick white stock. The flat surfaces are the parts that will be soaked and sealed. Using the required number of pieces, you will end up with the outer laminate, being another of the 1/4 in. white pieces. This should also result in a block approximately 5 in. x 5 in. x 3/4 in. thick. Allow to dry thoroughly.
4. Using disc sander, sand both flat surfaces.
5. Lay out hexagonal design on surface of the flat sanding.
6. On table saw or band saw, cut block to shape.
7. Sand and polish all surfaces.
8. Check size and thread of pen swivel to be used and drill and tap hole as required to take swivel plug.
9. Fit swivel plug into tapped hole in plastic base.
10. Screw swivel and funnel into place.
11. Clean and wax.

NOTE: The alternate design incorporates the same laminating process, however, instead of strip laminating with edges showing the four pieces of colored stock 4 in. square, are laminated face to face, alternately, and are cut to circle of 4 in. diameter, beveled and polished exactly like above strip laminating. Then steps 8, 9, 10, and 11 follow.

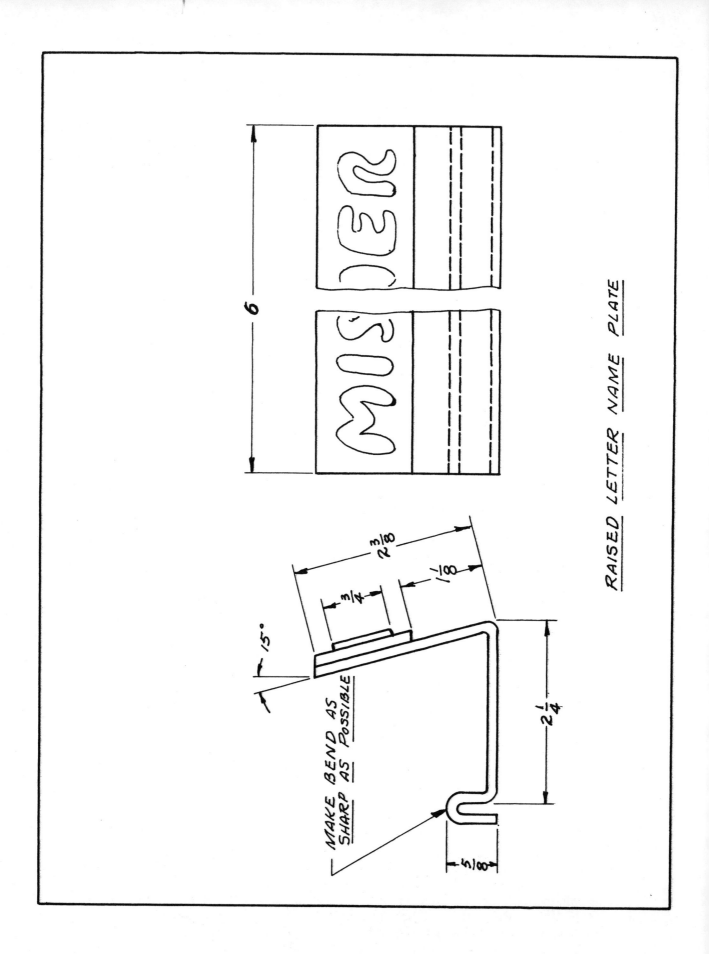

RAISED LETTER NAME PLATE

SUBMITTED BY:
Miss Alice Kuhlenkamp
Port Huron High School
Port Huron, Michigan

Here is a very unique name plate, simply made, yet made through quite an interesting process. The first impression is that individual letters have been cut out and cemented to the background. On closer inspection, one finds that the letters are a smooth blend of the material itself. In this case the formed base section is a separate piece from the name section, although it could be made as one unit.

MATERIAL REQUIRED:
1 piece white opaque plastic 1/16 x 1 x 6 in.
1 piece dark green plastic 1/8 x 1-1/4 x 6 in.
1 piece dark green plastic 1/8 x 6 x 6 in.

PROCEDURE:
1. Our primary concern is the making or obtaining of a design or lettering block to be used as the die. In making the name plate shown in the photo, the name or lettering die was made from a block of hardwood, carved out in reverse. The letters were about 3/4 in. high. The design or lettering die can be made of hardwood, or soft metal such as brass. If hardwood is used, care must be taken that the lettering is exact and that no splinters or broken edges appear, as every little defect will show up in the final product. Assuming a name plate die is at hand, we would proceed as follows.
2. Cut the two strips to size, remove masking paper from all parts, and solvent seal together. Allow to thoroughly dry, preferably in this case 10 to 12 hours, or overnight.
3. Place the cemented piece in oven at about 280 degrees F. for about 5 minutes or until thoroughly softened, remove, and quickly place white side against letter die and compress in vise. Compress lettering deep enough that white stock is below surface of green material. Allow to cool in this position.
4. When cooled, remove from vise, and carefully sand off entire white surface that remains above green material until only the white lettering remains, which will be even with the green surface. This should be the same as desired final contrast will appear. Sanding must be done gradually to prevent frictional heat from causing depressed letters to raise slightly, thus ruining or partially cutting away of letters in the process.
5. When sanded as desired, then wet-sand or smooth-sand this flat surface and all edges, and buff to high gloss.
6. Return to oven at same temperature, and allow to reheat. White lettering will return to original surface position and thus raised lettering will result, and will be fully polished.
7. Sand and polish edges of base section.
8. Heat and form base section in wood jig or by use of strip heater.
9. Solvent seal name plate to base section at desired level.
10. Clean and wax.

STAND For ELECTRIC CLOCK

Made from a single piece of clear plastic, this electric clock case or stand is both novel and practical.

MATERIAL REQUIRED:
1 piece clear plastic 1-3/4 in. thick x 4-3/4 x 6-3/4 in.
1 Electric clock movement (Lanshire) 3-3/8 in. size, or other appropriate movement
1 6-ft. white electric cord, plug, 2 connectors

PROCEDURE:
1. Cut block of clear plastic 1-3/4 in. x 4-3/4 in. x 6-3/4 in.
2. Using diagonal cross lines, lay out center of block for boring operation.
3. Place in 4-jaw chuck in lathe preparatory to boring hole for mounting of clock unit.
4. Bore hole to proper size to take clock movement.
5. Bevel edges on disc sander or in table saw set at 45 degrees.
6. Sand and polish all outer edges of block.
7. If internally carved design is desired, this should be done at this point prior to mounting of clock movement.
8. Clock movement is mounted in block, backplate secured in back, and cord attached.
9. Clean and wax.

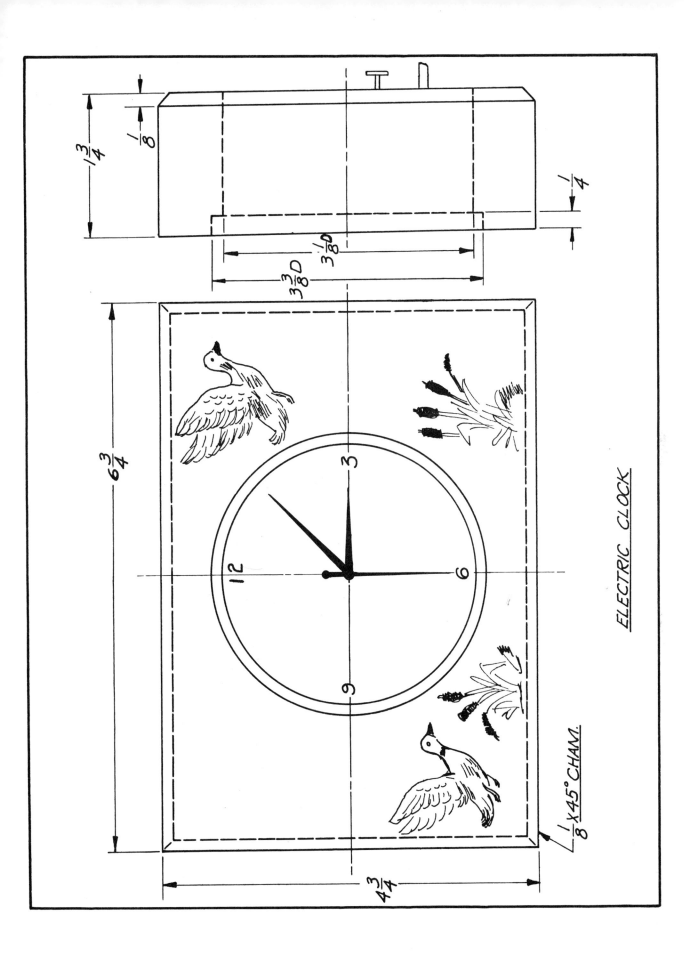

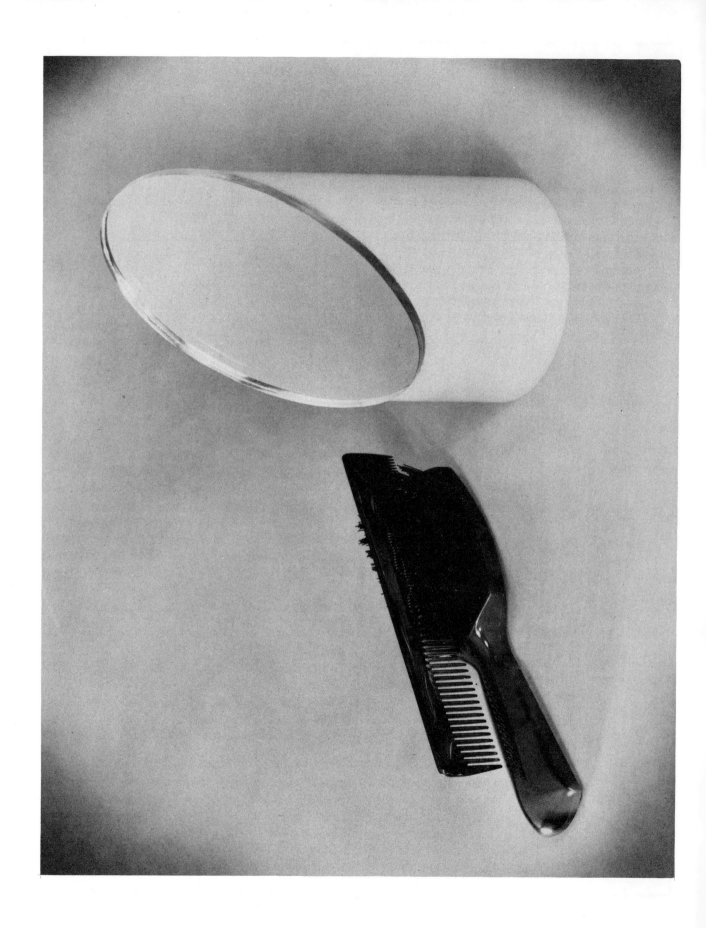

VANITY MIRROR

A clever new project made simply by cutting a piece of white translucent acrylic tubing at the proper angle, and sealing an oval of clear mirror acrylic into place.

MATERIAL REQUIRED:
1. The tubing is 4 in. OD x .100 wall white translucent acrylic tubing.
2. The mirror acrylic is 1/4 in. thick clear mirror acrylic stock cut into an oval shape to match the oval shape of the slanted tube cut.

The tubing can be cut to any length desired; in fact, sets of two or three varying sizes may be constructed for a "stair step" set.

PROCEDURE:
1. Cut the 4 in. OD tube to length, then cut the tube at desired angle (approximately 30 deg. cut).
2. Placing the angled cut of tube onto the mirror acrylic stock, trace oval outline onto masking. Cut the oval to size, and polish edges.
3. Using proper acrylic mirror stock adhesive, seal acrylic mirror oval to slanted cut of tube stock.
4. Square cut end of tube may be polished if desired.

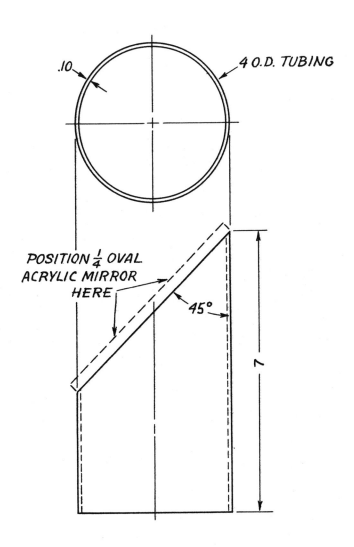

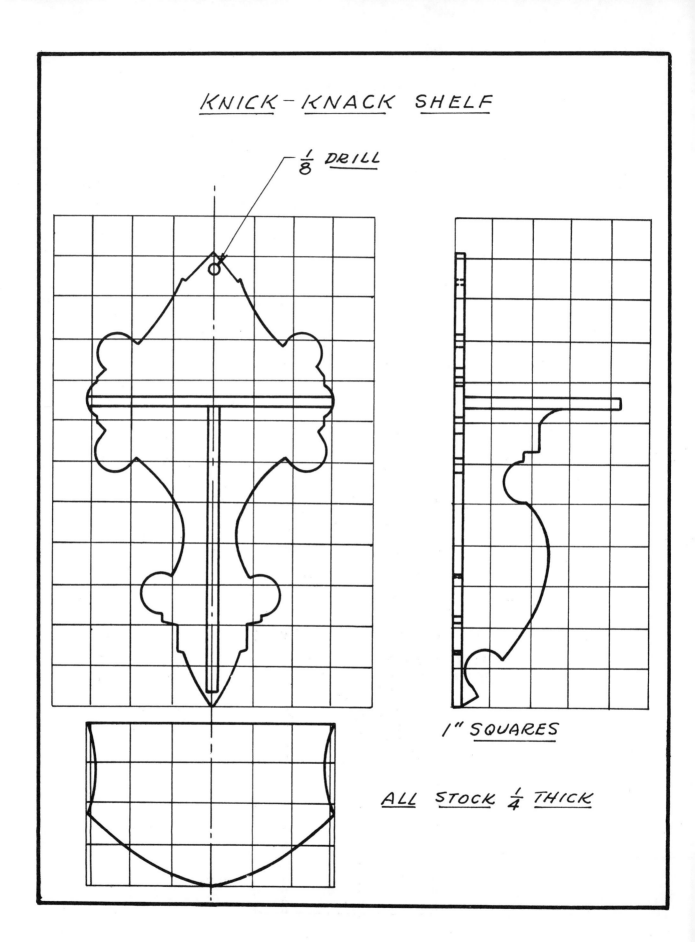

KNICK-KNACK SHELF

SUBMITTED BY:
Phil Brooks
University of Missouri
Columbia, Missouri

MATERIAL REQUIRED:
1 piece black plastic 1/4 x 4 x 6 in.
1 piece black plastic 1/4 x 3-1/8 x 7-1/4 in.
1 piece black plastic 1/4 x 6 x 11-1/8 in.

PROCEDURE:
1. Make templates for all three shaped panels using working drawings provided.
2. Using templates, trace all three designs onto masking paper of plastic.
3. Using band saw or scroll saw, cut out panels.
4. Sand all outside curvatures on disc sander when possible. Finish sanding by hand using #320 grit wet-or-dry sandpaper. NOTE: Rough saw cuts can be smoothed by using fine mill file prior to hand sanding. In some cases, buffing can be done directly from this fine filing.
5. After sanding operation, buff all edges to high luster.
6. Lay out hole location on masking paper and drill 1/8 in. hole through back panel as indicated.
7. Remove masking paper from all piecs.
8. Using soak method, place edge of center support member into Pleximent, then seal into position on back panel. Next soak edge of shelf member in Pleximent and seal in position above and adjoining center support. This may also be sealed to center support by running solvent along matching edge and surface after sealing into place on back plate.
9. Clean and wax.

HOT DISH PAD

SUBMITTED BY:
Harry Ruff
Coffeyville High School
Coffeyville, Kansas

This hot dish pad is a very unique project. It may be made up of any color of plastic desired, and laced with the plastic lacing of a contrasting color. The legs sealed to the bottom, may be of a color contrasting to the main body, or a close match in color to the lacing. The hot dish pad shown in the photo has black legs, yellow translucent body, and black lacing. A fine project at small cost!

MATERIAL REQUIRED:
1 piece yellow plastic 1/8 x 5-1/4 x 6 in.
1 piece black plastic 1/8 x 1-1/16 x 6 in.
3 yards black plastic lacing, 3/32 in. wide

PROCEDURE:
1. Lay out design for yellow plastic base; cut base panel to size.
2. From black plastic strip, cut out 6 leg units.
3. Sand and polish all outer edges.
4. Lay out hole placements and circular cutout, then drill 18 holes, 1/8 in. diameter for lacing. Drill hole in center section, preparatory to making jig saw cutout.
5. Cut out center section on jig or scroll saw, and sand or file smooth.
6. Remove all masking papers.
7. Using soak method, seal the six legs to base panel as shown, and allow to dry thoroughly.
8. Lacing Procedure:
Begin by placing the end of the cord down through hole 1, and glue this end to the underneath side. This leaves the long strip of cord on top. Now lace in following sequence: Take long end and go down through 10, up 5, down 14, up 9, down 18, up 13, down 4, up 17, down 8, up 3, down 12, up 7, down 16, up 11, down 2, up 15, down 6, then bring the end back to #1, and glue to the underneath side. Allow to dry. Holding the pad top side up, "down" would mean going from top to bottom side with the lacing; "up" would mean going from the bottom side to the top side with lacing.

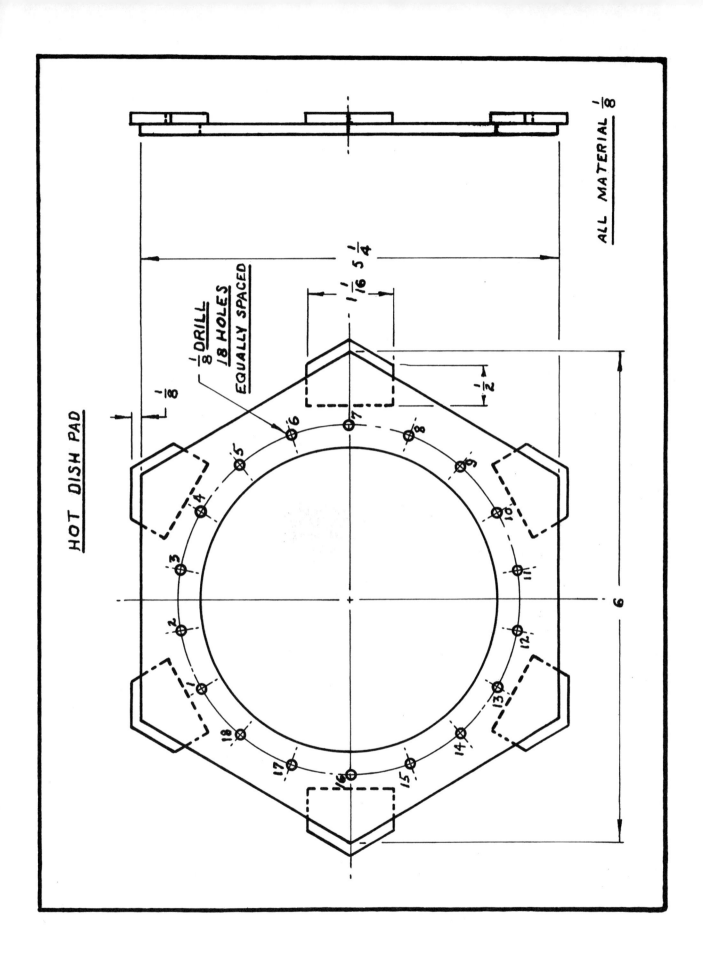

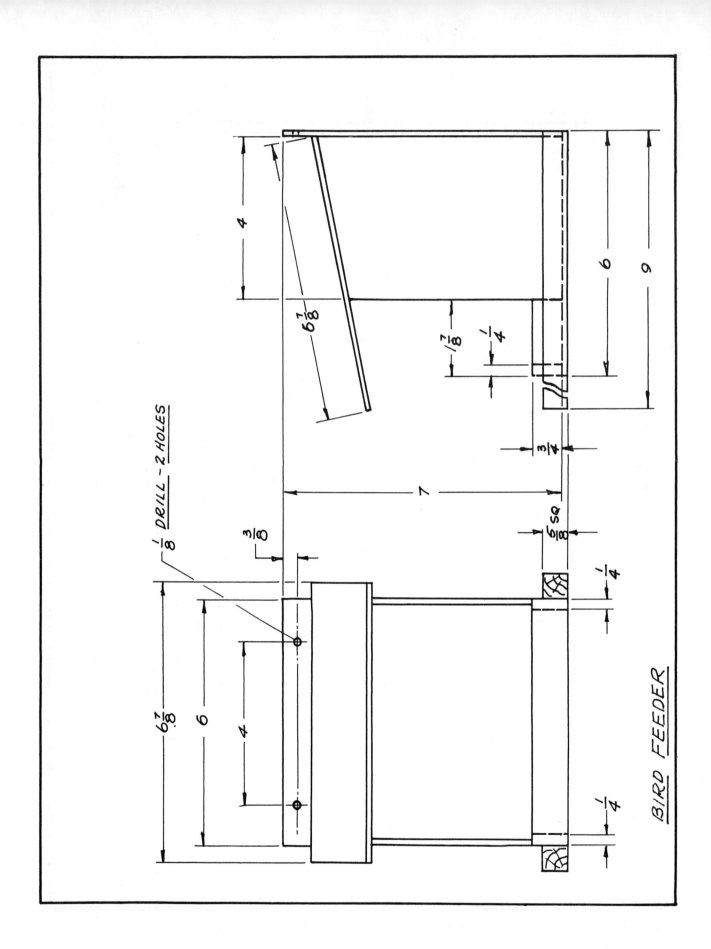

BIRD FEEDER

SUBMITTED BY:
Phil Brooks
University of Missouri
Columbia, Missouri

MATERIAL REQUIRED:
2 pieces black plastic 1/8 x 4 x 6 in.
1 piece black plastic 1/8 x 6 x 7 in.
1 piece white plastic 1/8 x 6 x 6 in.
1 piece yellow plastic 1/8 x 6-7/8 x 6-7/8 in.
1 piece clear plastic 1/4 x 3/4 x 6 in.
2 pieces clear plastic 1/4 x 3/4 x 1-5/8 in.
2 pieces wood 1/2 x 1/2 x 8-3/4 in.

PROCEDURE:
1. Cut plastic panels to size.
2. Sand and polish edges of materials as follows:
   (a) 2 long and 1 short edge of the 1/8 in. x 6 in. x 7 in. black panel.
   (b) The 1 edge measuring 5-5/16 in. of the two side panels (4 in. x 6 in. originals).
   (c) One long edge of each of the two pieces 1/4 in. x 3/4 in. x 1-5/8 in.
   (d) One long edge and both ends of the one piece 1/4 in. x 3/4 in. x 6 in.
   (e) All four edges of the yellow panel 1/8 in. x 6-7/8 in. x 6-7/8 in.
   (f) All four edges of the white panel 1/8 in. x 6 in. x 6 in.
3. Lay out and drill two 1/8 in. holes in black back panel for mounting screws.
4. Cut the two wood strips to size, sand smooth, and finish for outdoor use, using paint or spar varnish.
5. Seal the three black panels together, using soak method, thus forming back and two sides.
6. Sand top and bottom surfaces on disc sander, thus assuring flat surfaces for cementing to top and bottom.
7. Seal top yellow panel into position; then seal bottom panel into position.
8. Seal small side rail pieces to bottom plate.
9. Seal front rail to bottom plate, and at same time, to ends of side rails.
10. Glue the two wood perch rails into position, using #527 Bond cement.
11. Clean and wax.

STAND For RUBBER STAMPS

SUBMITTED BY:
Halemeyer Plastics
Golden Eagle, Illinois

Designed to take eight rubber stamps, this stand is a welcome addition to the home, the office, or school shop desk. The stand is designed so the top plate turns on the center pin, thus making any rubber stamp on the stand at your finger tips by just a slight turn.

MATERIAL REQUIRED:
1 piece clear plastic 1/4 x 6 x 6 in.
1 piece opaque white plastic 1/2 x 3-1/2 x 3-1/2 in.
1 piece opaque white plastic 1/2 x 1-1/2 x 2 in.
1 piece cast clear plastic rod 1-1/4 in. dia. x 2-1/2 in. long
1 piece clear cast plastic rod 1/2 in. dia. x 5/16 in. long
1 piece threaded brass rod, 10-24 thread, 1/2 in. long

PROCEDURE:
1. Cut the 1/2 in. white base piece to size.
2. Cut the 1-1/4 in. dia. clear rod to length, making sure both ends are cut square. Use lathe to turn small pivot knob on end of rod.
3. Drill and tap turned end of rod to take stud as shown in drawing.
4. Cut handle piece to shape and drill and tap 10-24 thread into small end (centered) as indicated.
5. Cut 1/4 in. thick circle to size, 6 in. dia.
6. Sand and polish all edges of all pieces made above except bottom end of 1-1/4 in. rod.
7. Seal 1-1/4 in. rod to base, centered.
8. Lay out and drill 3/4 in. holes in 1/4 in. thick stamp spindle as indicated.
9. Make cutouts from 3/4 in. holes to edge of spindle circumference.
10. Drill 1/2 in. hole through center of spindle.
11. Insert brass threaded stud in center post, assemble spindle, then screw handle down onto protruding rod.
12. Clean and wax.

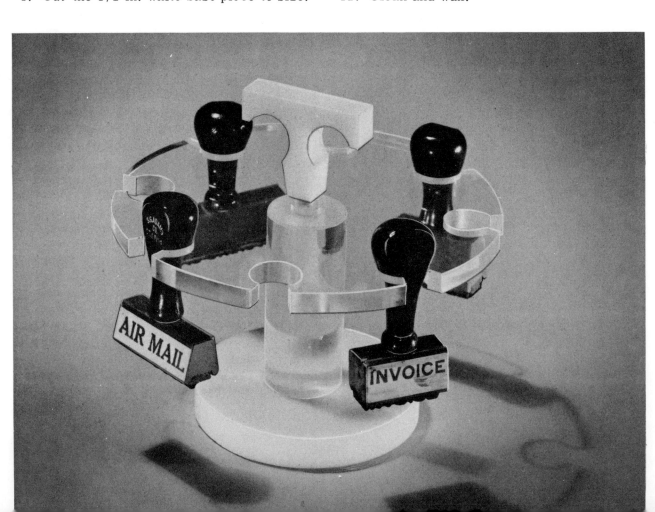

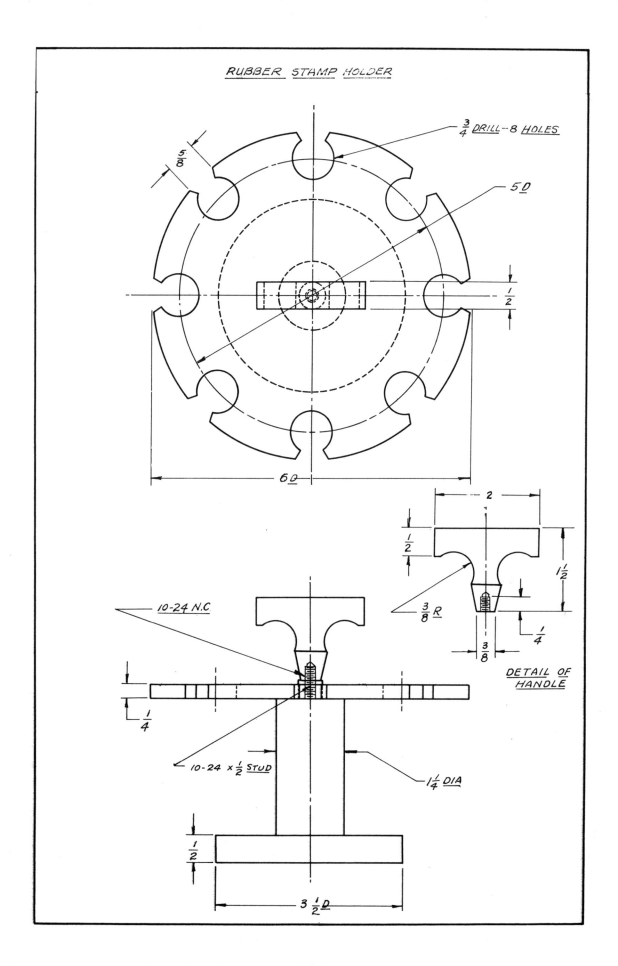

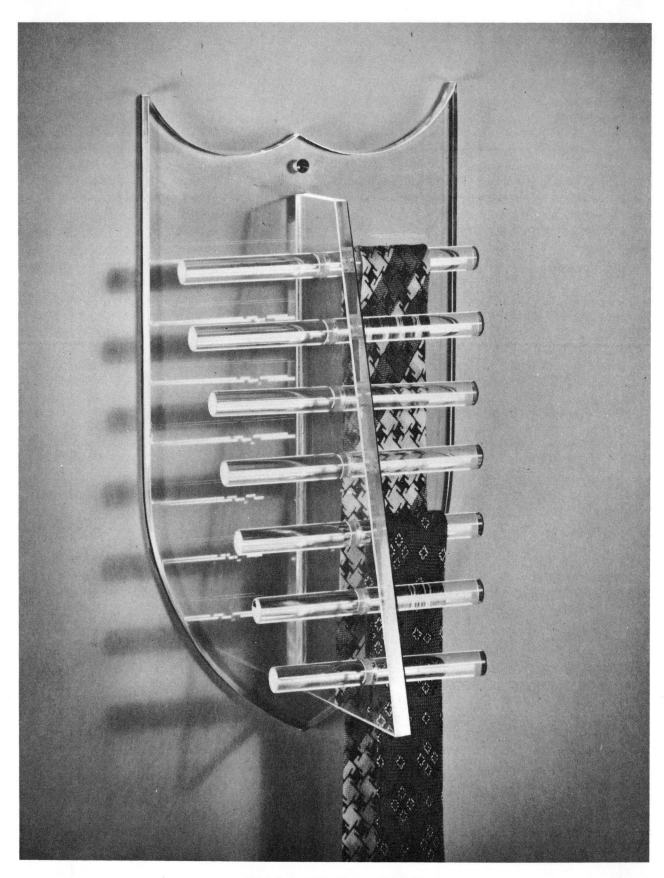

Tie Rack of Clear Plastic.

TIE RACK
MADE of CLEAR PLASTIC

SUBMITTED BY:
Phil Brooks
University of Missouri
Columbia, Missouri

MATERIAL REQUIRED:
1 piece clear plastic 1/4 x 5 x 9-3/4 in.
1 piece clear plastic 1/4 x 3-7/8 x 7-1/2 in.
1 piece 27 in. of 3/8 in. dia. cast clear plastic rod (enough material for sawcuts).

PROCEDURE:
1. To make template of back shield first make half pattern by drawing freehand design given in small squares into 1/4 in. squares on piece of heavy wrapping paper which has been folded. Cut while folded, then unfold to get full template.
2. Using template, trace this form onto masking paper of 1/4 in. x 5 in. x 9-3/4 in. piece of clear plastic.
3. Lay out center support section on masking paper of plastic piece 1/4 in. x 3-7/8 in. x 7-1/2 in.
4. Lay out hole locations.
5. Drill seven holes 3/8 in. diameter as shown. Drill one hole 1/8 in. diameter in back plate.
6. Cut out back panel and support panel as shown using band saw on back panel and either band saw or table saw for diagonal cut on support member.
7. Sand all edges. Buff to high luster. NOTE: Do not polish edge "A" which is to be sealed to back shield later.
8. From the piece of 27 inch 3/8 in. dia. cast rod, cut the following lengths: 1 each--4-3/4 in., 4-1/2 in., 4-1/4 in., 4 in., 3-3/4 in., 3-1/4 in., 3-1/2 in.
9. Sand and polish ends of rod. Ends may be rounded slightly for ease of insertion into support plate.
10. Using soak method seal support edge "A" to back shield as located.
11. Insert rods into support member, largest one at top etc. as indicated.

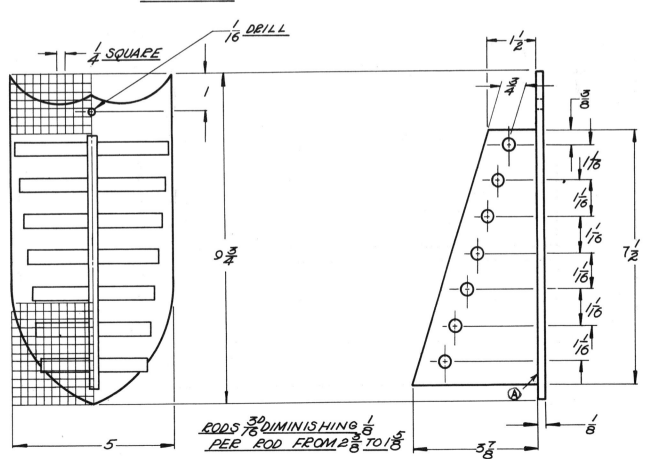

HEART-SHAPED PERFUME ATOMIZER STAND

SUBMITTED BY:
A. L. Kahler
Bloomsburg, Pennsylvania

This heart-shaped holder for a perfume atomizer vial is a colorful ornament that will enliven any girl's dressing table.

The heart-shaped body of the stand is laminated of red and white plastic. The photo shows a small cameo cemented in the center of the red face-piece of the heart. White initials or a white monogram would be equally appropriate.

MATERIAL REQUIRED:
1 block white plastic 3 x 3 x 1/2 in.
2 blocks red plastic 3 x 3 x 1/4 in.
1 piece red plastic 3 x 5 x 1/4 in.
1 piece white plastic 1-1/2 x 4-1/2 x 1/2 in.
2 pieces decorative white: small cameos, monogram or initial letter

PROCEDURE:
1. Remove masking paper from both sides of square 1/2 in. white block, and from one side of each of red squares. Clean unmasked surfaces with soft cloth dampened with 70 per cent isopropyl alcohol. Wet-sand surfaces. Bond red squares on each side of white square. Allow sufficient time (approx. 30 min.) for setting of bond.
2. Lay out heart shape on masking paper of one red face piece.
3. Saw heart shape from bonded red and white "sandwich" along line marked on

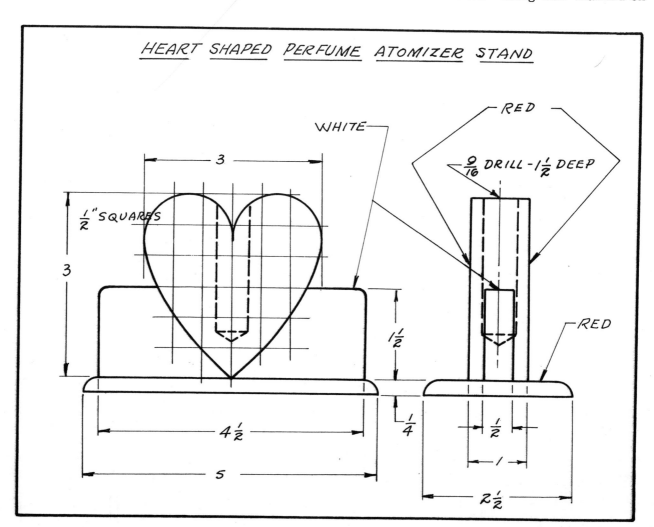

Heart-shaped perfume atomizer stand.

masking paper. Use jig or band saw, cutting through all layers at same time.

4. Drill hole down from lowest point on top of heart, clamping piece carefully to assure straight hole. Hole should be 9/16 in. diameter and 1-1/2 in. deep to accommodate perfume vial.
5. Remove masking paper from red faces of heart. Round off edges of red sheets with sanding disc, after smoothing heart contours on all sides of body piece.
6. On large piece of white plastic lay out outline of lower contours of heart-shaped body. Mark on masking paper, using already-formed heart-shaped body for pattern. Saw out cut using jig saw. Make cut true and smooth. Heart must fit exactly into this cut for bonding. Shape concave curves in white base piece to fit exactly against heart body by filing, or sanding. After curves are fitted, wet-sand (with motor-driven small drum sander, if possible) curved surface preparatory to bonding to body.
7. Wet-sand straight edges of white base piece. Be sure the two long edges (where white vertical base piece is bonded to red horizontal piece) are perfectly straight and square. Polish all sanded edges except where they are to be bonded.
8. Prepare red plastic base piece (3 in. x 5 in. x 1/4 in.). Round off edges on coarse sander, bring rounded part no more than 1/4 in. on upper face of red base piece, wet-sand and polish all edges.
9. Make suitable jig to hold vertical white base piece and heart body in position for cementing to each other and to red base piece. Add cement to edges of vertical 1/2 in. white base piece where it joins heart body and red base strip. Place parts in contact held in place by jig.
10. As soon as bond has set (allow at least 4 hours for proper setting) polish and buff all remaining edges. Cement on face decorative cameos, or other decorating. Insert atomizer vial in hole drilled in heart body.

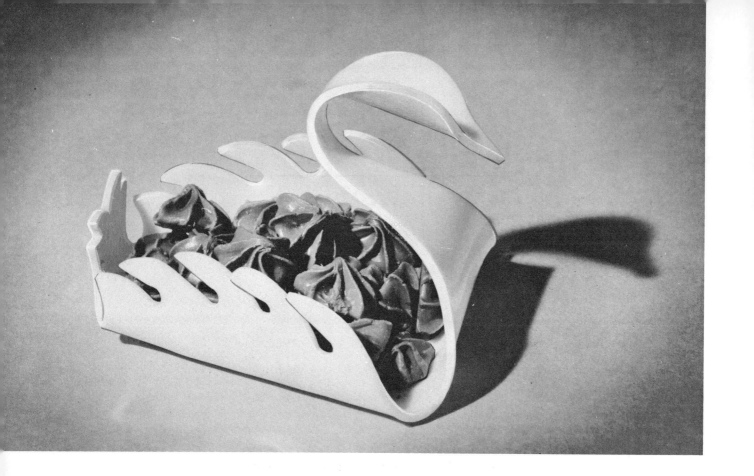

WHITE SWAN CANDY DISH

SUBMITTED BY:
Myrl Kirk
Enid High School
Enid, Oklahoma

To make this graceful candy dish of white plastic, plywood forms are used to accurately shape the plastic.

MATERIAL REQUIRED:
1 piece white plastic 1/8 x 6 x 12 in.
Approximately 1 sq. ft. of 3/4 in. plywood for forms

PROCEDURE:
1. The wood forms necessary to proper shaping of this candy dish are dimensioned by the drawings. Cut the form pieces, sand smooth, and nail into position. Six blocks are necessary, the main base block, the three pieces cut and glued permanently to the base, and the two blocks shown in side sketch. The two blocks shown at the side are used to press the material into place and are removed each time.
2. Make template for swan.
3. Cut piece of plastic 1/8 in. x 6 in. x 12 in.
4. Using template, draw pattern off onto the paper masking on plastic.
5. Using scroll saw, make swan cutout.
6. Using fine rat tail file and abrasive paper, smooth all surfaces.
7. Polish edges to high luster.
8. Remove masking paper.
9. With oven set at about 275 deg. F. heat panel for about 4 minutes until flexible.
10. With the aid of a pair of cotton gloves, remove plastic and place over form. Using block A as pressure block, press soft plastic down into center area of form. This will make flat bottom, and will force edges of swan body up into upright position along blocks C, D, & E.
11. Bring block B into tail position, forcing edge of tail section, into upright position.

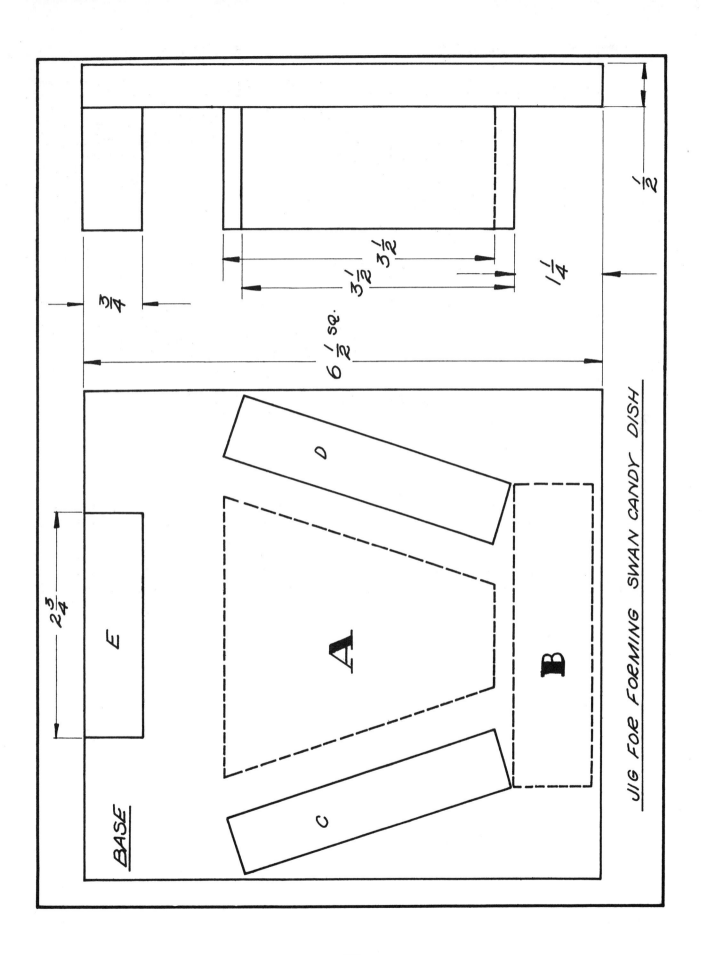

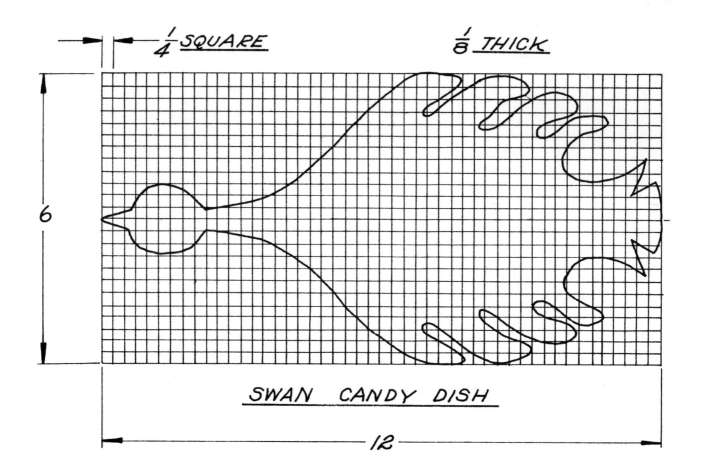

12. Place weights on two removable blocks to hold them in place.
13. The swan neck and head are formed and then held in place by hand until cool. A forefinger of one hand can be used to shape the neck area. The curvature of upsweep in the head is formed by using the forefinger, thumb, and second finger of other hand to hold it into position until cool. Because this material cools rather quickly and sets, the forming operation should be done as quickly as possible.

PLANTER LAMP

SUBMITTED BY:
Frank Dohanich
Frankfort High School
West Frankfort, Illinois

This lamp incorporates the use of a glass brick along with the sparkling clear plastic to become a Planter Lamp. The rich black base along with the polished aluminum (or brass if preferred) rods makes a striking contrast and a very beautiful lamp.

MATERIAL REQUIRED:
1 piece clear plastic 3 x 3 x 10 in.
1 piece black plastic 1/2 x 3 x 12 in.
1 piece black plastic 1 x 1-1/4 x 3 in.
2 polished aluminum (or brass) rods, 1/4 in. dia. x 15 in. long
4 #10-24 flat head bolts 1 in. long
1 glass brick 4 in. x 7-3/4 in. (standard)
1 piece threaded lamp pipe 3/8 x 1 in. long
1 lamp socket (with screw on base connection to fit pipe)
1 lamp bulb
1 6 ft. cord and plug
1 shade (your selection)

PROCEDURE:
1. Cut the pieces of plastic to size. (1) 3 x 3 x 10 in. clear, (1) 1/2 x 3 x 12 in. black

173

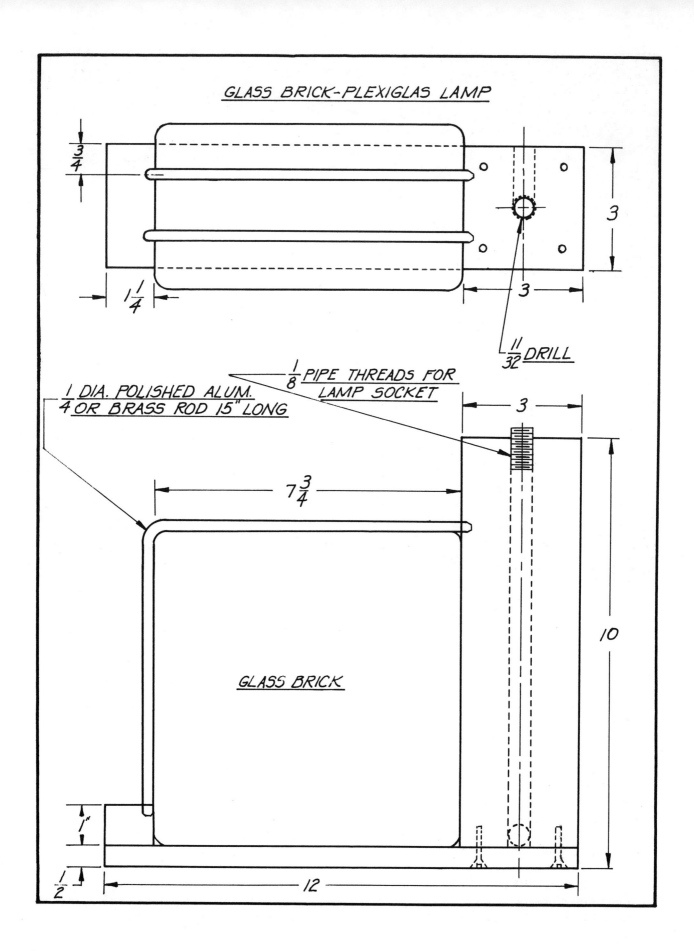

Projects From Plastics

    (1) 1 x 1-1/4 x 3 in.
2. Sand and polish all three pieces.
3. Locate center of 3 in. bar from end, and drill 11/32 in. hole through entire length. This should be drilled from one end to assure full alignment of hole through block. (An 8 in. piece of 5/16 in. steel rod can be welded to your present 11/32 in. drill to serve as extension). Drill or rout groove in one end of 3 in. bar from center hole to one edge for cord entry.
4. Locate and countersink, then drill clearance holes through black bottom plate for bolt insertion. Drill and thread matching holes about 3/4 in. deep into one end of 3 in. bar, to receive 10-24 bolts.
5. Dip the short pipe tube into Pleximent and thread into 3 in. bar on opposite end from routed groove and bolt holes. Pleximent, solvent, will soften hole surface enough to allow pipe to cut its own threads. Repeated dipping and threading may be needed to cut threads to desired depth which should be about 1/2 in. into bar.
6. After pipe nipple is in place, it is desirable to block off that end of bar, and pour white enamel from opposite end into the long drilled hole. Then pour out and allow to dry overnight. This will result in a pure white gleaming drilled hole surface which in turn will hide white cord to be used. The enameling is optional.
7. Remove masking papers.
8. Locate and drill two 1/4 in. diameter holes as indicated in 1 in. black block for later receiving of polished rods.
9. Seal black block in position as indicated.
10. Place glass brick in position merely to determine bend of rods.
11. Place end of polished rod into hole in black block, determine exact point of bend to be made around end of glass block, then bend rod over end and along adjoining edge. Repeat with second rod. This may be done in vise after determining exact bend point.
12. Place 3 in. block in position and determine exact location of two holes to be drilled to receive rod ends. Rods should enter 3 in. bar about 1/4 in. deep. If too long, cut rods to proper length.
13. Drill two holes in 3 in. bar as located (1/4 in. diameter x 1/4 in. deep).
14. Thread the 6 ft. cord through bar, then place bar in position and bolt securely. This will anchor the glass brick in place.
15. Screw socket onto pipe nipple, and make cord connection.
16. Clean and wax.

SWEPT-CORNER BOX

SUBMITTED BY:
V. S. Fox
Lincoln High
Cleveland, Ohio

Intended for costume jewelry, this box is unique in the design of its sweeping lines that flow smoothly in graceful curves. Added to this is the brilliant contrast of the deep red transparent base and lid enclosing the stark white sides of white opaque plastic.

The knob on the lid is made from a lamination of red transparent material faced on both sides with thin strips of opaque white.

MATERIAL REQUIRED:
2 pieces, transparent dark red plastic, 9 x 2-3/4 x 1/4 in.
2 pieces, opaque white plastic, 9-1/8 x 2-5/8 x 1/8 in.
1 piece, red transparent plastic, 2-5/8 x 1/2 x 1/4 in.
2 pieces, white opaque plastic, 2-1/4 x 1/2 x 1/16 in.

PROCEDURE:
1. First bend sides of box to shape indicated by drawing. Remove masking paper from both white strips, clean, and prepare for bending. It is essential that both side pieces be given exactly the same curvature, since this is one of the beautiful and graceful features of the box. It is suggested the side strip be heated over a small hot plate or other small source of heat, so the heat is concentrated about one-fourth of the distance from end of strip, rather than heating the entire strip in an oven.
2. After the strip has been heated to the softening point, it should be quickly bent to the required curve, since 1/8 in. plastic cools quickly at room temperature. If a jig has been provided, place heated strip in jig to cool. If a jig is not used, one piece may be bent by hand, allowed to cool, and used as a guide for bending the second side strip. Be sure to put a piece of thin, soft cloth between the two strips to prevent sticking and mark-off.
3. Carefully sand bevels on end of side strip nearest curved part, so this end will fit exactly against the long side of the other strip. Fit two strips together to form sides of box, and cement at two edges of contact.
4. Sand off protruding edge of long end of each side to bring edge into outline of curve of the adjoining edge, making a sharp, single-line corner at each long end of the box.
5. After box sides have been firmly cemented together and set, sand both top and bottom edges of box until it sits flat and true when placed on a flat surface.
6. Cement sides of box to bottom piece.

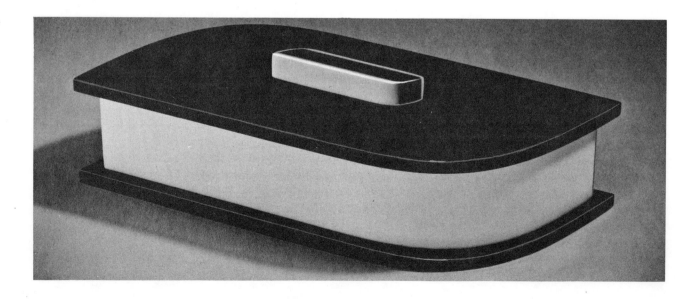

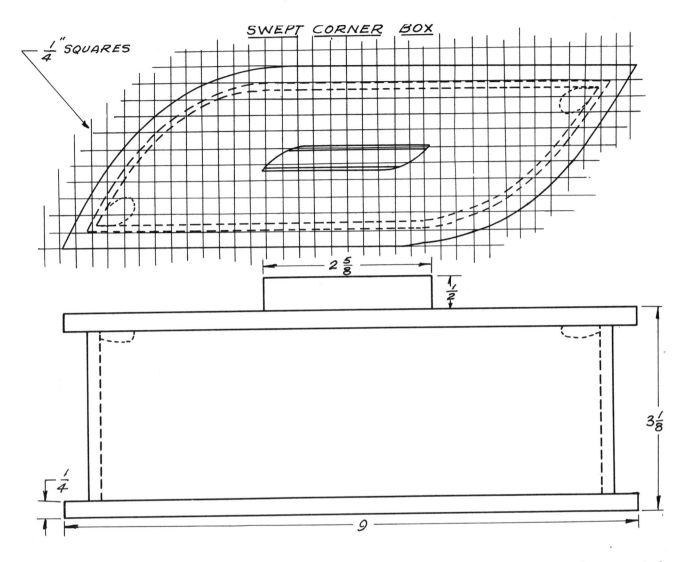

Keep the long, straight portion of the side exactly 1/4 in. in from the edge of the red bottom piece.

7. After sides have been cemented to bottom piece, saw curved edges on bottom piece, taking care to maintain the 1/4 in. border all the way around the box.
8. Sand and polish edges of bottom piece.
9. Using finished bottom of box for pattern, saw top from other large piece of 1/4 in. red plastic. Sand and polish edges.
10. Prepare laminated piece for knob, cementing the two 1/16 in. pieces of white material on the two faces of the 1/4 in. red piece.
11. Sand curve on ends of knob to correspond with curve of lid. Sand "flat" on one edge of knob piece where it is to be cemented to lid. Polish all other surfaces.
12. Prepare two "buttons" of red plastic 1/16 in. thick in teardrop shape to serve as stays for lid. These may be cut from 1/4 in. stock. Cement these to under side of lid so that point of teardrop piece is in exact corner of box. It is possible, by careful work, to slightly trim these pieces after they have been cemented in place. They should be so adjusted that when they are forced down into corners of box by putting lid in place, they will hold lid in position and offer enough frictional contact so lid will not be easily loosened.
13. Clean all surfaces, wax, and polish with soft cloth.

Three Tier Tray.

THREE TIER TRAY

SUBMITTED BY:
Ernest Belden
Carthage High School
Carthage, Missouri

The three curved trays in this unique project are shaped over wood forms turned on a wood lathe.

MATERIAL REQUIRED:
1 piece black plastic 1/8 x 8 x 8 in.
1 piece black plastic 1/8 x 4 x 4 in.
1 piece white plastic 1/8 x 6 x 6 in.
1 piece 5/8 in. diameter cast plastic rod 8 in. long

PROCEDURE:
1. Lay out and cut to size the three plastic discs.
2. Sand and polish all edges; then remove masking paper.
3. Place one plastic disc in oven at about 275 deg. F. When pliable remove and form to curve of turned block. Repeat for other two discs using proper forms. Disc must be held in position until cooled sufficiently to retain its shape.
4. Drill holes to take 5/8 in. rod, in centers of 6 in. and 8 in. trays.
5. Solvent seal end of rod in hole in 8 in. tray. Allow to dry.
6. Fit 6 in. tray on rod in proper place and solvent seal.
7. Cement 4 in. tray onto upper end of rod and allow to dry.
8. Clean and wax.

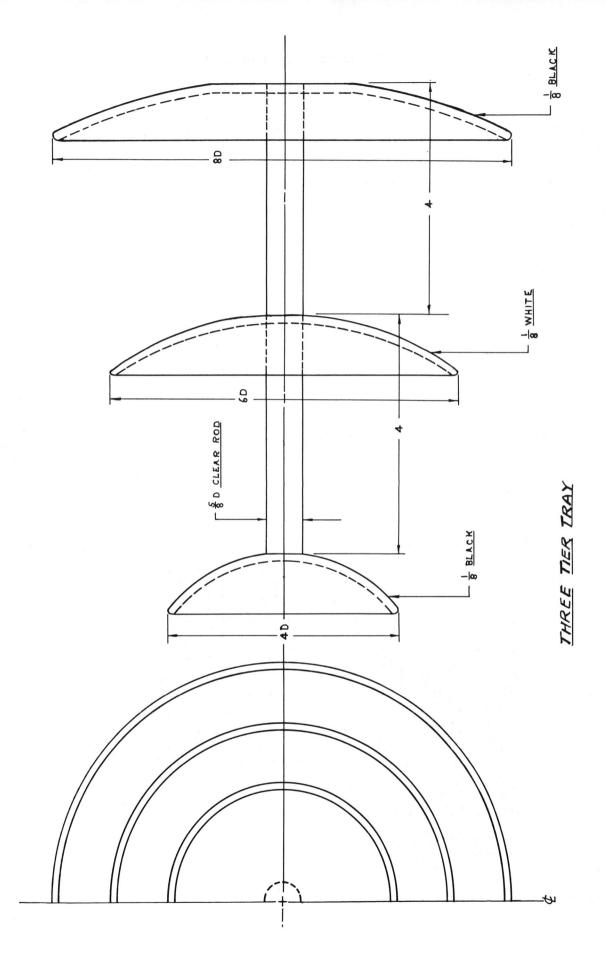

SALT and PEPPER SHAKERS

SUBMITTED BY:
Harry Ruff
Coffeyville High
Coffeyville, Kansas

These salt and pepper shakers are the result of heat forming by use of the strip heater. The black sides of the main barrel are formed from one strip which has been bent at three places and formed to the box shaped sides, then sealed at the one edge. The contrasting color of the top panel and the monogram make it a useful and outstanding item. Base is provided with a 1/2-20 threaded plug which serves as the filling hole, and when threaded into place makes a completed seal.

MATERIAL REQUIRED:
2 pieces black opaque plastic 1/8 x 2-1/2 x 6-1/4 in.
2 pieces black opaque plastic 1/4 x 2 x 2 in.
2 pieces ivory (or white) plastic 1/8 x 1-1/2 x 1-1/2 in.
2 threaded plugs 1/2 in. - 20 x 1/4 in. long
About 4 sq. in. of matching ivory plastic 1/8 in. thick for monograms

PROCEDURE:
1. Saw the 2-1/2 in. x 6-1/4 in. pieces of black plastic to size and mark off the exact location of the corners to be bent. See drawing.
2. Set rip fence on saw and with blade raised about 1/16 in. make saw kerfs across panel. This will aid in making square corners for shaker barrel.
3. Cut base and top panels to size, and round off corners to approximate 1/8 in. radius.

Projects From Plastics

Bevel bases at 45 degrees per drawing. Bevel or round off top panels to pleasing contour.

4. Sand and polish top and base pieces.
5. From scraps of ivory (or white) plastic cut monograms of the letters "S" and "P" (for salt and pepper), and sand and polish edges.
6. Lay out holes in top, and drill nine holes 1/16 in. diameter in each top.
7. Lay out center of base pieces, and drill and tap with 1/2-20 thread.
8. Remove masking paper from all parts.
9. Plug in strip heater, and when it has reached full temperature, place the black plastic panel with saw kerfs cut into it over heater (with kerfs directly over element--but not touching it), and after two or three minutes, first fold can be made. Hold at 90 degree angle, until cool, then continue with second bend, and finally the third bend. Then cement final joint, thus, forming box shape.
10. If any material extends out over final joint, sand this off even with the rest of the surface, and round corner to same radius as has been bent with other three. Sand and polish.
11. After box shaped barrel has thoroughly dried, touch both open ends to disc sander to make them absolutely flat.
12. Using soak method, seal barrel to base, then seal barrel to lid in same manner.
13. Seal monograms to location desired; usually centered.
14. Insert 1/2-20 plugs in base (screw into position flush with bottom).
15. Clean and wax.

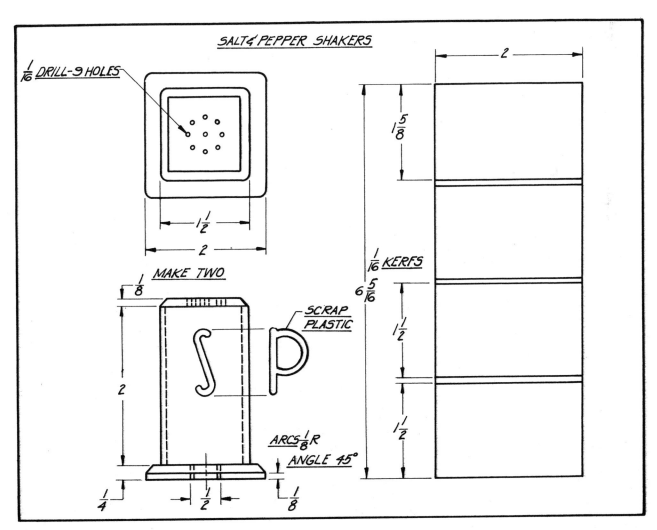

Cattail Centerpiece.

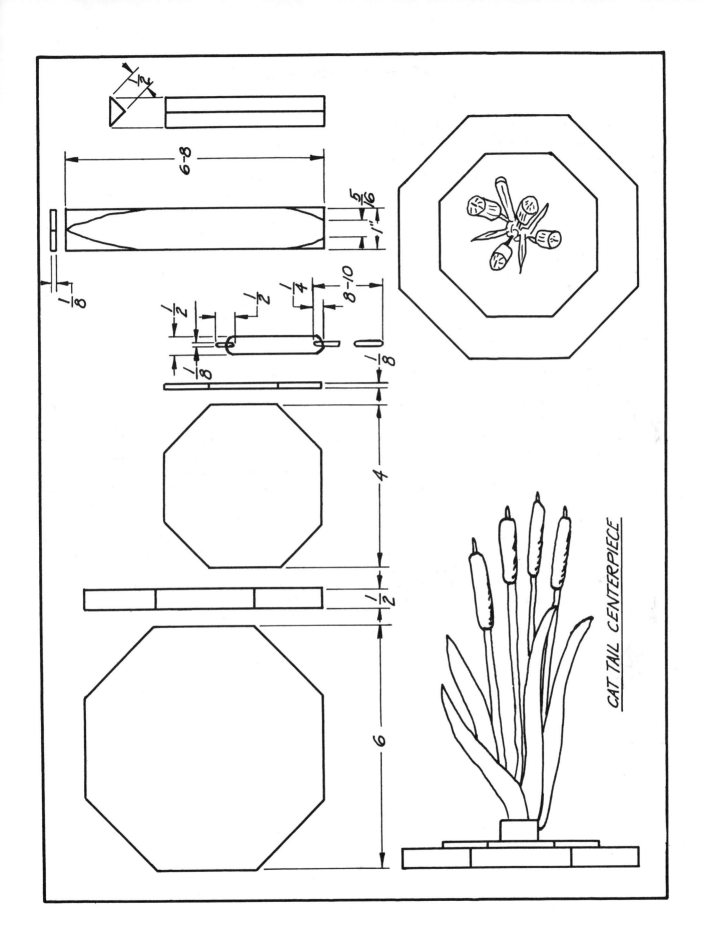

CATTAIL CENTERPIECE

This centerpiece simulates a cattail plant. The white, frosty cattails on the ends of the slender stems of clear plastic rod, the dark green, shining leaves, all mounted on a white and green octagonal base make a very colorful piece.

MATERIAL REQUIRED:
1 piece white plastic 6 x 6 x 1/2 in. (base)
1 piece dark green plastic 4 x 4 x 1/8 in. (upper base)
4 pieces clear 1/8 in. plastic rod, 8-10 in. long (stems)
4 pieces clear plastic rod 1/2 in. dia. 1-3/4 in. to 3 in. long
4 pieces 1/8 in. plastic rod 1/2 in. long (cattail tips)
4 pieces dark green plastic 1/8 in. thick x 3/4 in. to 1 in. wide--6 to 8 in. long (leaves)
1 strip triangular cross-section clear plastic about 4 in. long, made by sawing off corner of 1/2 in. piece of plastic along length of side, with saw set at 45 degrees.

PROCEDURE:
1. Prepare octagon-shaped piece from 1/2 in. white plastic square by sawing off corners of square. Bevel and polish edges.
2. Make octagon from 1/8 in. x 4 in. green square. Polish edges and bevel upper corner.
3. Bond green octagon to center of white octagon.
4. Cut stems from clear 1/8 in. plastic rod in lengths from 8 in. to 10 in.
5. Drill 1/8 in. dia. hole, 1/4 in. deep, into each end of the four pieces of 1/2 in. rod. Round off ends of these cattail body pieces by rough sanding. Also sand side surfaces to "frost" body piece. Cement 1/2 in. piece of 1/8 in. rod into one end of each cattail body piece, and rough sand this to "frost" it.
6. Insert "stems" of 1/8 in. clear plastic rod into end of cattail bodies, bonding them in place.
7. Drill four holes in base to receive stems. Bond in place.
8. Prepare leaves by sawing pointed end on each of four green strips and polish edges.
9. Heat leaf strips in oven and bend into random patterns, as shown in photo.
10. Prepare four supporting strips of clear plastic that has been cut at 45 degree angle--making supporting triangular strip slightly shorter than leaf is wide. Bond one perpendicular edge of this strip to base of leaf, then bond leaf and supporting piece to green base surface, surrounding stems of cattails already in place.
11. Clean and wax.

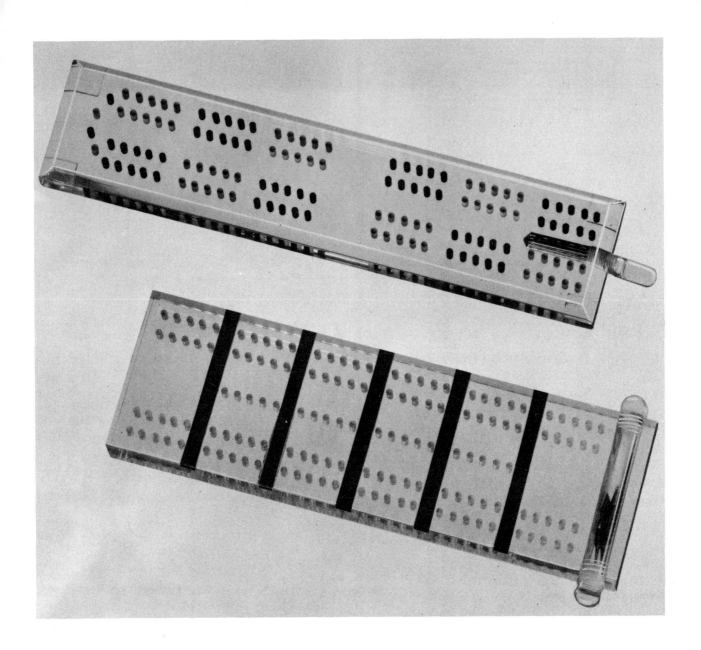

CRIBBAGE BOARD

SUBMITTED BY:
William Lantz, Industrial Arts Instructor
Marysville, Kansas, and
Lloyd Gouine, Industrial Arts Instructor
Cheboygan, Michigan

Two cribbage boards are pictured here. We are using the one with the inlaid stripes (left) as our example for fabrication.

MATERIAL REQUIRED:
1 piece clear plastic 3/4 x 3 x 9-3/4 in.
1 piece blue plastic 3/16 x 1/4 x 12 in.
(allows for sawcuts and pegs)
1 piece red plastic 3/16 x 1/4 x 12 in.
(allows for sawcuts and pegs)
1 piece 1/2 in. dia. clear cast plastic rod 2 in. long

PROCEDURE:
1. Saw the clear panel to size, 3/4 in. x 3 in. x 9-3/4 in.
2. Saw five pieces of blue plastic 3/16 in. x 1/4 in. x 1-9/16 in.
3. Saw five pieces of red plastic 3/16 in. x 1/4 in. x 1-9/16 in.

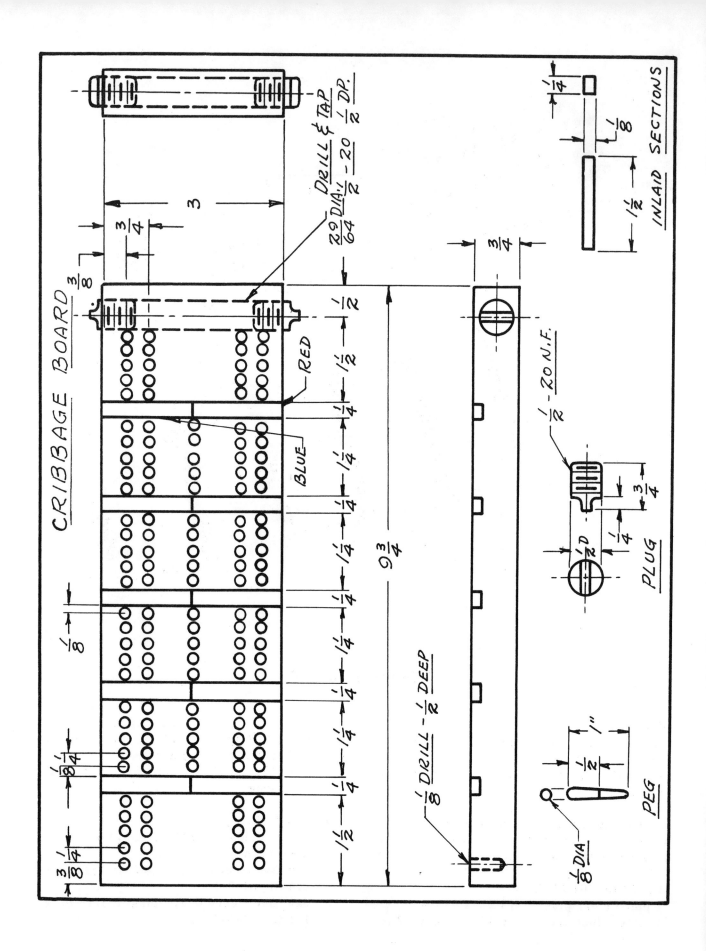

4. From drawings, lay out hole positioning of peg holes and also make lay out of grooves to be cut for inlaid pieces of red and blue.
5. Raise circular saw blade 3/16 in. above table top, and using guide and fence, saw grooves to receive inlaid strips. These grooves should be 1/4 in. wide or grooved to fit the inlay pieces already cut to size. These should be a good snug fit.
6. Drill 140 holes 1/8 in. diameter to 1/2 in. depth, as laid out in drawing. Drill carefully and accurately because these holes will look bad if not properly aligned. It is well to use a piece of plastic with a straightedge clamped to the drill table, to serve as a guide in aligning a series of holes running the same way.
7. Place the cut and fitted strips of red and blue plastic in the solvent, a couple at a time and allow the surface to soften for about one minute. Insert a blue strip up to the half way mark, and a red strip opposite it, thus, the two meet and fuse in the center. Since these strips were cut to 1-9/16 in. at the beginning, you will have a bit left out of the groove. The excess is removed by sanding. Continue to insert the strips and seal until all five sets are in place.
8. Sand and polish the plastic panel.
9. Remove paper masking from plastic.
10. On edge, 1/2 in. from one end, and centered of the 3/4 in. thickness lay off a center for drilling of the cross hole for peg storage. Using a 29/64 in. drill, drill hole through plastic, using coolant if necessary to obtain smooth finish hole.
11. Using 1/2-20 steel tap, tap the hole with this size thread for about 1/2 in. (or slightly over) depth, from both ends.
12. The 1-5/8 in. piece of clear rod 1/2 in. diameter is then threaded on each end with the use of a threading die to a 1/2-20 thread. This thread should extend about 1/2 in. from both ends. Upon completion of the threaded ends, cut the piece exactly in half, and you will end up with two pieces of rod approximately 3/4 in. long, each with a threaded end. These two plugs then may be polished off flat on the surface, or a grip flange cut or filed onto the end opposite the threads. In either case, it is then polished on the grip end.
13. Using the remaining portion of the 12 in. piece of 3/16 in. red and blue stock, cut three pegs of each color, and lathe turn or hand file to shape as shown in drawing. These, of course, serve as the pegs for the cribbage board, and are stored in the hole in the end of the main block.
14. Clean and wax all parts.

NOTE: As an alternate to drilling the large end hole in the plastic block all the way through, you may just wish to drill part way through, thus producing a blind hole, and thus requiring only one threaded plug. Also, this hole may be made into the endwise positon as pictured in the alternate design.

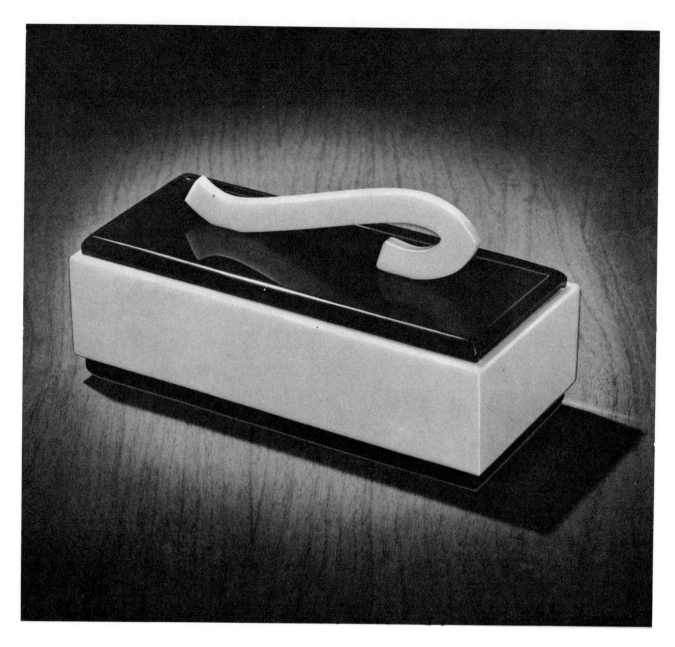

BLUE and WHITE CANDY BOX

SUBMITTED BY:
V. S. Fox
Lincoln High School
Cleveland, Ohio

This box which is somewhat larger than plastic boxes described previously, has sides of opaque white plastic, with bottom and top of deep blue transparent material. The blue lid is set off by a white handle cut in pleasing curves.

Dimensions of the box are 8-1/2 in. x 3-13/16 in. x 3-3/4 in.

MATERIAL REQUIRED:
2 pieces transparent blue plastic 8-1/4 x 3-1/2 x 3/8 in.
2 pieces opaque white plastic 8-1/2x2x1/4 in.
2 pieces opaque white plastic 3-13/16 x 2 x 1/4 in.

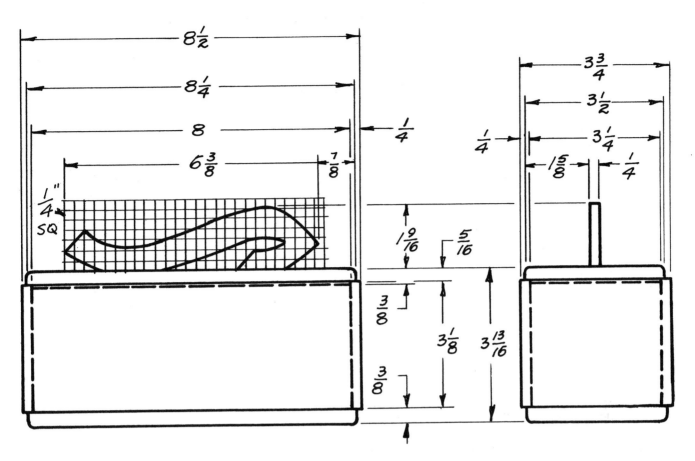

BLUE AND WHITE CANDY BOX

1 piece opaque white plastic 6-3/8 x 1-9/16 x 1/4 in. (For handle)

PROCEDURE:
1. Remove the masking paper from one side of each of the four pieces of opaque 1/4 in. plastic.
2. Cut a 45 degree bevel or mitre on each of the narrow ends of these four pieces. This can best be done by sanding on a mechanical sanding disc equipped with a table that has been tilted to 45 degrees.
3. Cement mitered corners together. It is best to use a jig for this job to make sure the corners are square and the sides parallel. Leave the cemented sides of the box in the jig until they are well cured.
4. Remove masking paper from blue top and bottom pieces. Round off upper corners of lid and lower corners of bottom plate on sander, making radius of curvature equal to thickness of piece (3/8 in.). Polish all edges.
5. Place cemented box frame on belt sander or other flat sanding surface, and sand upper and lower edge of box until all are flat and lie in same plane.
6. Cement blue bottom piece to bottom of box, using solvent-soak method.
7. Make saw kerf along edges on bottom side of lid, cutting away just enough material so about 1/16 in. of lid fits snugly down inside inner edges of top of box sides. This is to hold lid in place.
8. Buff and polish upper edges of box sides. These have been sanded in step (5).
9. Saw handle piece from remaining strip of 1/4 in. white plastic. Lay out pattern from drawing on upper masking paper and saw piece with scroll or jig saw. Sand all edges. Sand "flats" as shown in drawing thus giving two fairly large bonding areas where handle can be bonded to top of lid. Buff and polish edges except on "flats." Cement handle to lid.
10. Clean and wax.

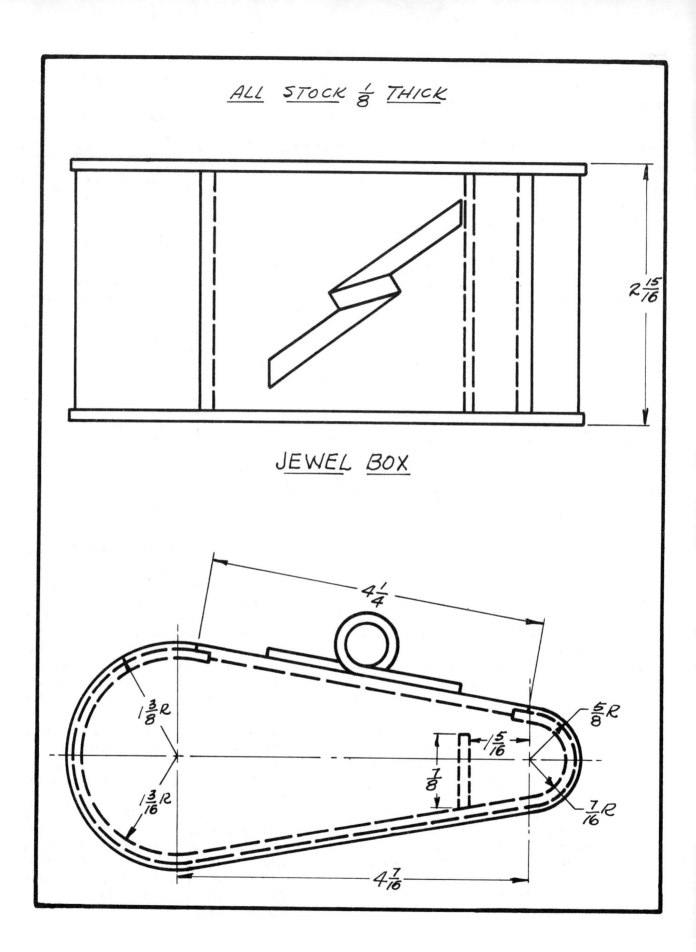

BLACK and WHITE JEWEL BOX

SUBMITTED BY:
V. S. Fox
Lincoln High School
Cleveland, Ohio

This box is really a jewel. The sides overlap the bent piece that forms the ends and bottom and curve around to form part of the top. The side pieces and lid are made of opaque white plastic. The curved piece and handle are of black plastic. The total effect of the stark white and the gleaming black is one of contrast that adds much to the attractiveness of the unusual design.

MATERIAL REQUIRED:
- 2 pieces opaque white plastic 6-1/2 x 2-1/2 x 1/8 in.
- 1 piece opaque black plastic 10-7/8 x 2-15/16 x 1/8 in.
- 1 piece black plastic 2-15/16 x 7/8 x 1/8 in. (Cross brace)
- 1 piece white plastic 4-1/2 x 2-15/16 x 1/8 in. (Lid)
- 1 piece black plastic 5-3/4 x 1/4 x 1/8 in. (Handle)

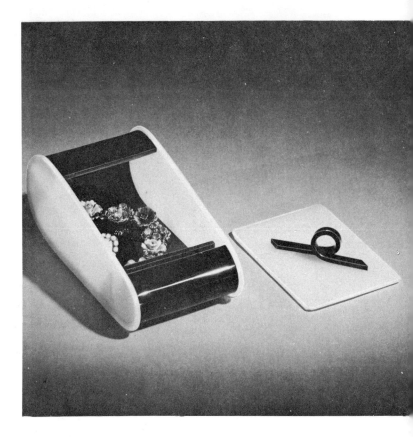

PROCEDURE:
1. Make a jig (of a block of wood, for molding the bent piece). The jig has rounded ends and is 2-15/16 in. wide and 6-1/16 in. long. The ends are curved to radii as shown in the drawing.
2. Remove masking paper from long piece of black plastic, and after heating, form about jig block to shape, as shown in drawing.
3. After piece has cooled and set, sand both edges until truly flat on belt sander or sandpaper laid on flat surface. This is to assure perfect fit to flat sides.
4. Cement flat side pieces of white plastic to sides of box, then saw to correct contour--1/16 in. away from outer surface of black, curved body piece. Sand and polish edges of white side pieces.
5. Cement black strip 2-15/16 in. x 7/8 in. x 1/8 in. in place in front part of box, 1-3/16 in. back from inner contour, as shown in diagram. This piece acts as a separator, and also strengthens front end of box.
6. Prepare lid by cutting saw kerfs successively in top and bottom edges. The kerfs are about 1/16 in. deep, and are cut successively back from bottom edge until lid fits down precisely into hole in top of box. After finishing saw kerfs, round off four corners slightly by sanding, then sand and polish edges.
7. Prepare handle by center-heating 1/4 in. wide strip black plastic. Form loop in handle by wrapping soft piece about 1/2 in. wood dowel. While still soft, press two flaring ends of handle onto flat surface.
8. Polish edges of handle except surfaces to be in contact with lid. These should be sanded lightly by holding flat surfaces on belt sander for short period.
9. Bond handle to lid, placing it diagonally across lid and somewhat above center, to make it more nearly in center of completed box.
10. Wax and polish with soft cloth.

Table lamp.

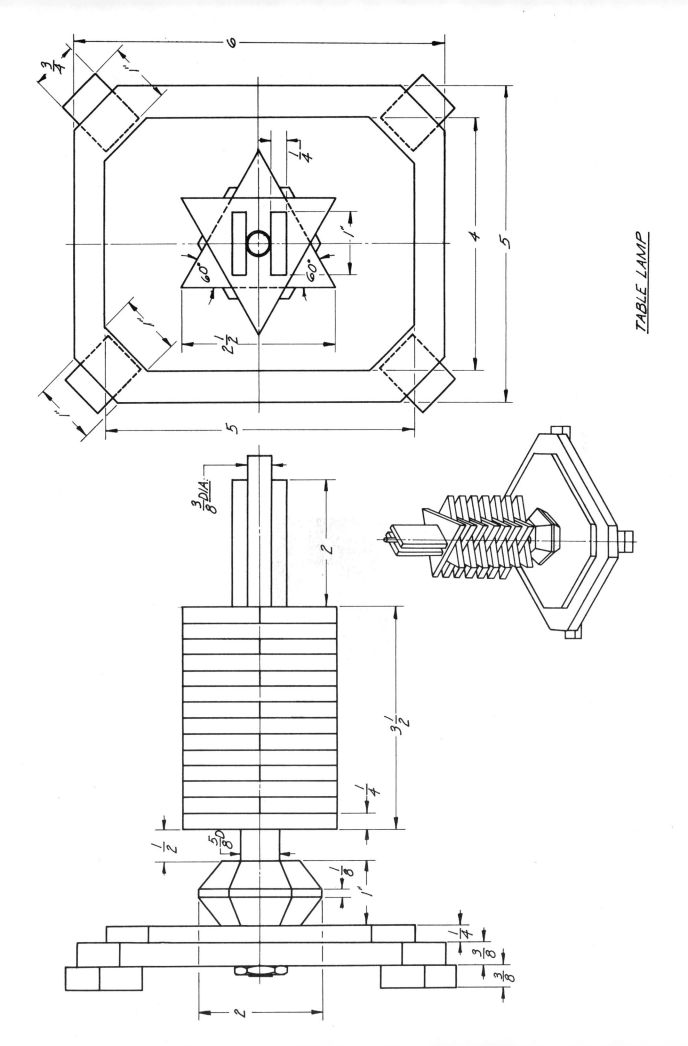

TABLE LAMP

SUBMITTED BY:
Walter Ambrose
Hadley Technical High
St. Louis, Missouri

In making this lamp of plastics, scrap material may be used for the small clear triangular pieces, thus making the total cost very moderate. Only two colors of plastic are used, rich black and highly polished crystal clear plastic.

MATERIAL REQUIRED:
1 piece clear plastic 3/8 x 5 x 6 in.
1 piece black plastic 1/4 x 4 x 5 in.
14 pieces clear plastic 1/4 x 2-1/2 x 2-1/2 in.
4 pieces clear plastic 3/8 x 3/4 x 1 in.
2 pieces black plastic 1/4 x 1 x 2 in.
1 piece clear plastic 1 x 2 x 2 in.
1 piece threaded lamp tube (metal) approx. 9 in. long
1 electric socket threaded for lamp tube
1 only 6 ft. cord and plug
1 lamp shade
1 brass spacer 1/2 in. long x 5/8 in. OD x 1/2 in. ID

PROCEDURE:
1. Cut two base panels to size; cut corners at 45 degrees as indicated.
2. Sand and polish edges of above two plates.
3. Seal black plate to clear plate, centering so 1/2 in. edge border remains all around. Allow to dry.
4. Locate center of this laminated plate, and drill 25/64 in. hole through both pieces.
5. Cut small 3/8 in. x 3/4 in. x 1 in. "feet" blocks (4), sand, and polish.
6. Seal these blocks at corner of base as indicated (to clear piece).
7. Cut 1 in. x 2 in. x 2 in. block to size, and bevel at indicated angles to octagon shaped piece. Sand and polish. (This block may be any shaped block that brings out the reflective qualities of the polished plastic--square, hexagon, octagon, or any other attractive shape). Locate center and drill 25/64 in. hole.
8. Cut and polish 14 of the 1/4 in. thick triangles per drawing. Locate center points of each and drill 25/64 in. hole.
9. Cut and polish two 1/4 in. x 1 in. x 2 in. black plastic blocks.
10. Begin assembly by inserting brass threaded tube through base, and screw locknut onto end of tube. Then slide 1 in. thick decorative block over top of tube end and down against base. Next, slide brass spacer tube into position. Follow this up with the 14 clear triangles staggering them alternately to form pattern indicated.
11. Seal two black decorative strips to top clear triangle close to the tube as indicated. Allow to dry thoroughly.
12. Screw socket onto tube end and lock in place. Tube if too long may be cut off with hack saw so approximately 3/8 in. extends above black strips at top. Tighten locknut on bottom, thus drawing all parts together.
13. Run cord end under base, up through tube, and connect to lamp socket.
14. Place bulb of desired wattage in socket.
15. Clean and wax.

CRADLED JEWEL BOX

SUBMITTED BY:
V. S. Fox
Lincoln High School
Cleveland, Ohio

This attractive jewel box consists of a framework of clear red plastic, which supports a cradled, semicylindrical box of translucent white plastic, which swings on pins of clear plastic rod. The box has a hinged lid of translucent white plastic, hinged by cementing to one end a 3/16 in. rod which passes through holes in the outer supporting frame.

MATERIAL REQUIRED:
- 2 pieces red plastic 1/4 x 2-3/8 x 5-1/16 in. (Frame sides)
- 1 piece red plastic 1-15/16 x 3-1/2 x 1/8 in. (Rear cross-piece frame)
- 1 piece red plastic 1 x 3-1/2 x 1/8 in. (Front cross-piece, frame)
- 1 piece white translucent plastic 3-1/2 x 4-1/4 x 1/8 in. (Lid)
- 1 piece white translucent plastic 6 x 3-1/8 x 1/8 in. (to heat-form into semicylindrical shape for cradle)
- 2 pieces white translucent plastic 3-1/2 x 2 x 1/8 in. (ends of cylindrical cradlepiece)
- 1 knob
- 1 piece 3/16 in. clear plastic rod 5-1/2 in. long

PROCEDURE:
1. Cut score lines in 6 in. piece of 1/8 in. translucent white material. Make first line 1/2 in. in from edge on ends and 5/16 in. in from long edges. Second, make parallel lines 1/4 in. in from each of first lines.
2. Heat 6 in. white piece to forming temperature and bend into semicylindrical shape, using cardboard mailing tube, or similar cylinder, 3-1/4 in. outside diameter. Take care not to get a twist into this piece when bending. Tie to forming tube with soft twine until cool.
3. When cool, sand curved ends on belt sander until perfectly flat.
4. Cement 3-1/2 in. x 2 in. x 1/8 in. pieces to two open curved ends of above piece to form sides of container. When this cementing job has cured, saw end pieces carefully around curved line. Then sand these edges, with slight bevel to make smooth fit along curved edges of both ends

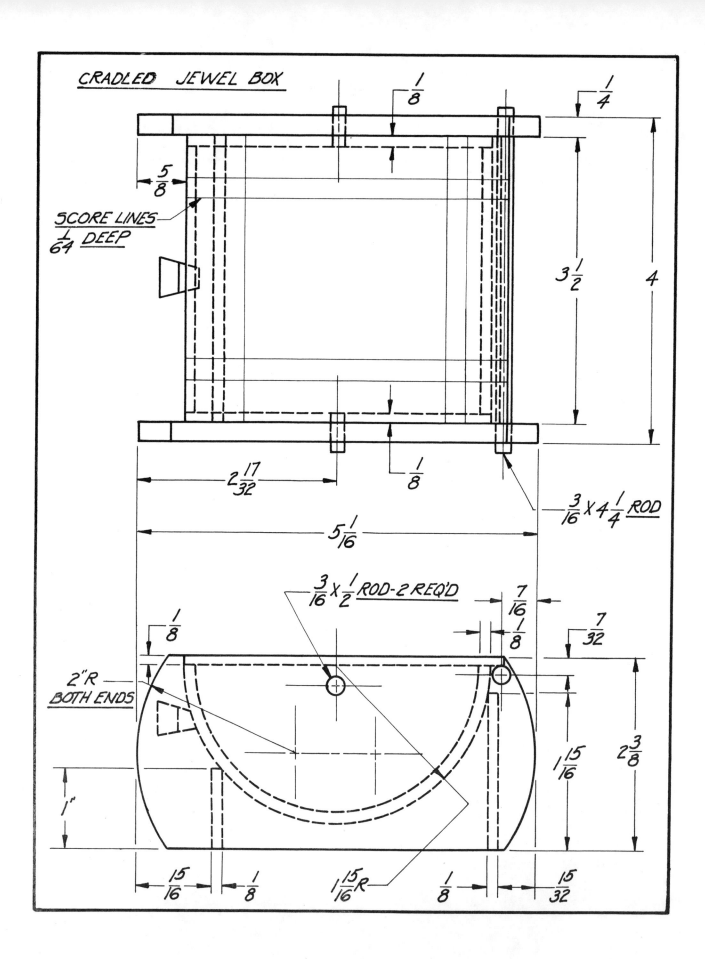

of cradle-like box. Polish all edges.
5. Make red plastic frame. Cut two 1/4 in. side pieces to shape as shown in drawing, making curved ends of each piece to same arc of curvature. Polish all edges.
6. Cement two 1/8 in. red strips to sides to form cross-pieces of red frame. Make sure that bottoms of these cross strips are even with bottoms of red side pieces; and that they are proper distance apart (front to back) to allow curved cradle enough room to be swiveled on its hinge rods. Polish top and bottom edges before cementing.
7. Mark location of holes for swivel rods in semicircular ends of cradle. Center hole 1/4 in. down from top edge, and in exact mid-point (front to back) of end pieces. Drill 3/16 in. holes.
8. Mark location of corresponding holes in 1/4 in. red side pieces. These holes should center 7/16 in. from top of red side piece, and should be located (front to back) by marking through holes already drilled in ends of cradle, making sure cradle will have room to rotate completely between front and back strips of red frame.
9. Cut two 1/2 in. lengths of the 3/16 in. rod Pass these through corresponding holes in ends of cradle and side pieces. After making sure that everything fits, remove these rods, "soak" 1/8 in. of end of each rod in cement and reinsert into these holes, dry end first, and working from inside of cradle, so that when rods are seated the soaked ends will be in the white translucent end pieces of cradle. This will cement these rods to cradle while leaving them free to rotate in holes in red sides of frame.
10. Make box lid from remaining pieces of 1/8 in. white translucent plastic. Score decorative saw cuts in under side of lid. Line up these score marks with score marks in round body piece of cradle.
11. Sand and polish edges of lid. Take remaining piece of 3/16 in. rod (which should be about 4-1/8 in. long) and fit through holes drilled in upper rear corner of red side pieces of frame. These holes should be 3/16 in. diameter, and centered about 1/4 in. down from top edge of red side piece. When this rod is in place, lid piece (prepared in (10) above) should fit down exactly on rod and also on top of semicylindrical cradle portion of box. Cement lid to rod along back edge of lid, about 1/4 in. in from edge--or in correct position so box lid lies symmetrically in top of frame when closed.
12. Attach knob to center, top front edge of cradle. Knob in photo shown in truncated pyramid cut from block made by color-cementing two pieces of 1/4 in. clear plastic, then cutting in square pyramid section that is 7/16 in. square at base and 1/4 in. square at point of truncation. Cement knob at small end to front of cradle.
13. Wax and give final polish with soft cloth.

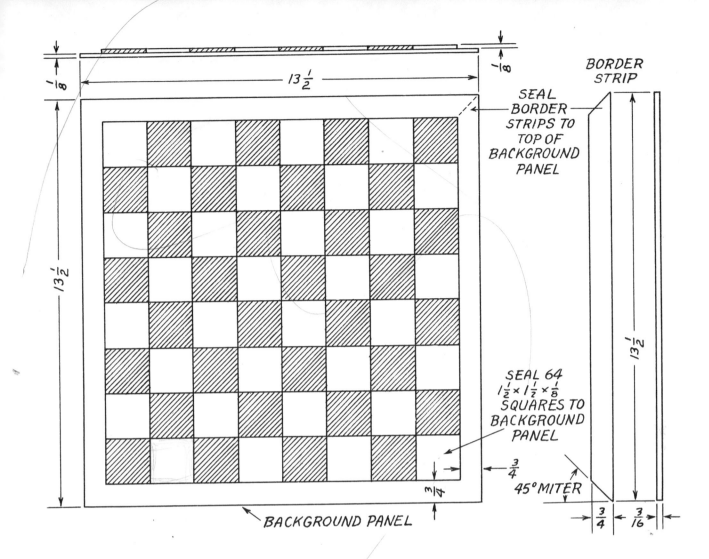

CHECKERBOARD

A beautiful Plexiglas checkerboard may be constructed by using two contrasting colors of Plexiglas throughout.

MATERIAL REQUIRED:
- 1 piece 1/8 x 13 1/2 x 13 1/3 in. ivory acrylic plastic.
- 4 pieces 3/16 x 3/4 x 13 1/2 in. ivory acrylic plastic.
- 16 pieces 1/8 x 1 1/2 x 1 1/2 in. ivory acrylic plastic.
- 16 pieces 1/8 x 1 1/2 x 1 1/2 in. black acrylic plastic.

PROCEDURE:
1. Cut the 13 1/2 in. square ivory background to size. Do not polish edge at this point.
2. Cut the 32 pieces 1 1/2 in. ivory and black acrylic pieces to exact size, and be sure they are perfectly square.
3. Cut the 4 pieces of 3/16 in. ivory side strips to size, and miter both ends at 45 deg. as shown.
4. Layout positioning of the 1 1/2 in. squares onto the background piece, placing the colors alternately in the group. Be sure the pieces fit squarely and accurately onto the background, then start sealing them down, working one row at a time in perfect alignment.
5. After all squares are sealed down, place border strips in position and seal up tightly to the rows of squares.
6. If outer edges of the unit do not match perfectly at this point, trim outer edges at one time. Sand and polish outer edges.

Plastic checkers may be made by cutting 3/8 in. sections of plastic rod while chess pieces may be turned or carved from scrap plastic stock.

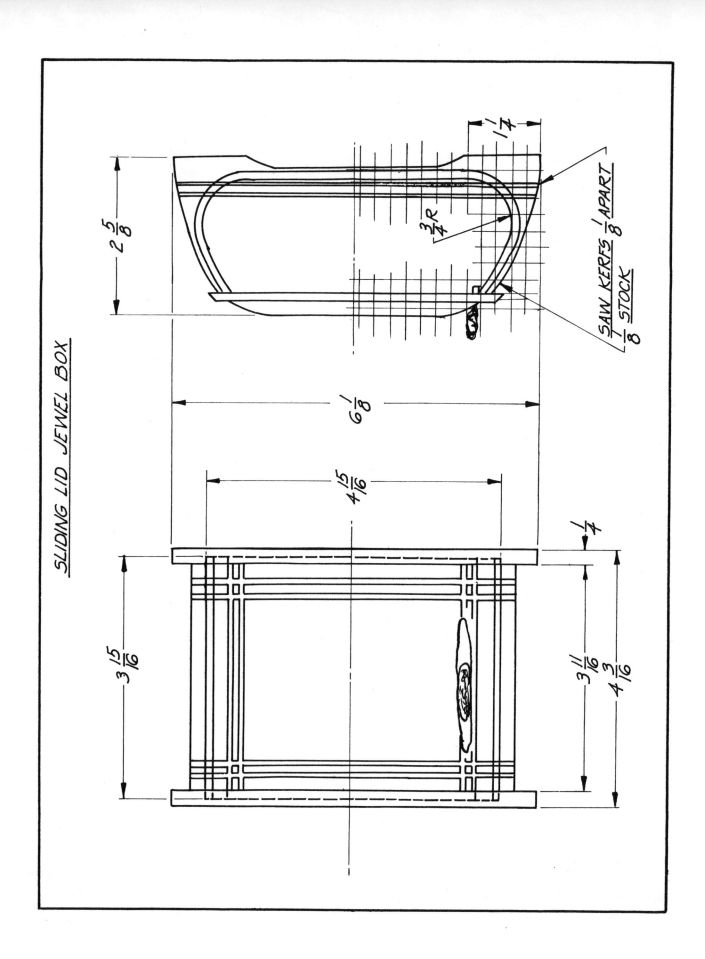

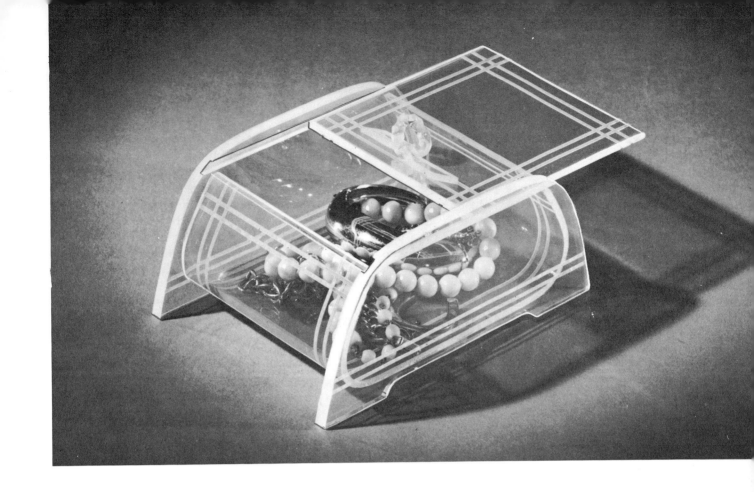

SLIDING LID JEWEL BOX

SUBMITTED BY:
V. S. Fox
Lincoln High School
Cleveland, Ohio

This jewel box with its sliding lid, made of transparent, red fluorescent plastic is a "glamour" item for lady's dressing table.

MATERIAL REQUIRED:
2 pieces 1/4 in. red fluorescent plastic, 2-5/8 x 6-1/8 in. (ends of box)
1 sheet, same material, 1/8 in. thick x 3-3/4 in. wide x 9 in. long
1 sheet, 1/8 x 4-15/16 x 3-15/16 in. same material, for lid
1 strip, 1/8 x 3/16 x 4-1/2 in. for making twisted knob
1 piece, 1/8 x 1/8 x 3/8 in., for stop

PROCEDURE:
1. Make parallel decorative saw kerfs parallel to all four edges of the 3-3/4 in x 9 in. sheet, spacing kerfs 1/4 in. in from edge, and 1/8 in. apart.
2. Heat 9 in. long sheet in rather wide band centering about 1-7/8 in. from end of sheet by resting it some distance above heater element of strip heater and bend as shown in drawing--do this on each end of the sheet, and bring bent strip to approximate dimensions shown in drawing.
3. When this piece is cool and rigid, place two edges (those that will form the front and rear rim of the box) on belt sander, or other flat surface sander, and sand off corners of edges so they will make a straight line over which the lid can slide.
4. Flat-sand the other two open ends of this bent piece so that each will continuously touch a flat surface.
5. Saw two 1/4 in. end pieces to identical shapes, as shown in drawing. Cut out 3/16 in. deep recess along bottom edges of these pieces, beginning and ending cuts about 1-1/4 in. from each tip of the bot-

tom edge. Also saw groove near top of each piece, slightly wider than 1/8 in., to provide grooves for lid. These grooves should be parallel to bottom edge, and the lower edge of each groove should be even with the top edge of curved section, both front and back top edges of box.

6. Cement ends of box to curved piece that forms front, bottom, and back. The box should be assembled on a flat surface so it will stand square when finished.

7. Prepare lid from 1/8 x 4-15/16 x 3-15/16 in. piece. Cut decorative saw kerf marks. Bevel front and back edges of lid back as shown.

8. Prepare knob by heating 4-1/2 in. long narrow strip and twisting it about six twists, holding the ends untwisted for about 3/4 in. back from each end. While still hot, bend the twisted portion into a ring, and bring out the untwisted ends of the strip in opposite directions and with the wider face perpendicular to the plane of the ring. Be sure that both flat faces of these ends are in the same plane so the knob can be easily cemented to the lid.

9. Cement stop piece to under side of lid at exact point where it will allow the lid, when closed, to protrude the same distance, front and back, past the top edges of the box.

10. Clean and wax.

LAPEL PINS ARE EASY TO MAKE

Procedure

1. Trace design to be used onto transparent paper.
2. Fasten paper to colored plastic--1/4 or 3/16 in., using rubber cement.
3. Saw out design with hand or power saw. Remove paper pattern.
4. Cement pinback to back of cutout, as shown in detail.

COCKER SPANIEL   SCOTTIE   BOSTON BULL   PIN BACK

THREE-CONE TABLE LAMP

SUBMITTED BY:
A. LeRoy Kahler
Bloomsburg, Pennsylvania

This striking, decorative lamp has that "something different" look about it that is eye-catching and distinctive, yet the piece is relatively easy to construct. The base and cone-shaped lamp holders are made from white, opaque plastic. Each cone contains a small electric bulb fitted into a candelabra base--the bulbs used are clear bulbs of the type used in 120-volt Christmas tree lights. The base of the lamp shown in the photo is made from a patterned plastic, such as is sometimes available--however, the usual smooth type may be used equally as well. The cones of the lamp shown in the photograph are 6, 5 and 4 inches high. The picture shown at the right is a block of carved Plexiglas, though any decorative piece of suitable size-- a small, framed photograph, may be used.

MATERIAL REQUIRED:
1 piece of 1/4 in. white plastic, about 6 x 10-1/2 in. (base)
3 pieces, 1/8 in. white plastic sheet about 8 in. long and ranging from 6 to 4 in. wide (cones)
3 triangular pieces, 1/4 in. white plastic (supports for cones). These may be scrap left from sawing off corners in making base plate
9 pieces of 1/8 in. clear plastic rod, three 2-1/2 in., three 3 in., and three 3-1/2 in. long
9 "plastic stones" -- small decorative pieces that can be obtained from a plastics supply house. Suggest black or amber color for suitable contrast. These are used on tops of pistil-like protrusions from the cones
3 one-half inch plastic balls, to be used for "feet" for the lamp
Set of electrical findings, including three

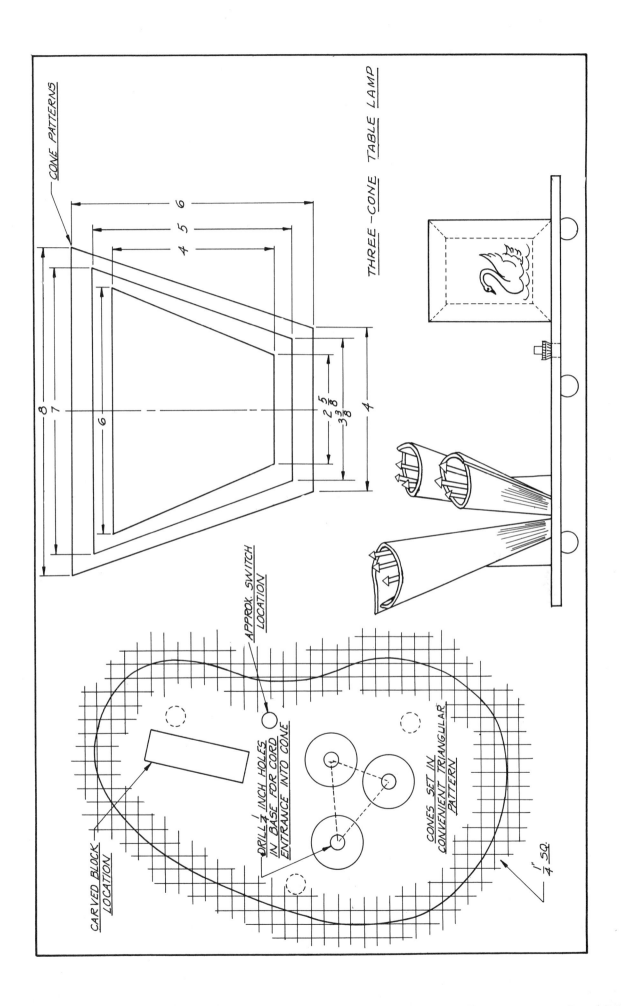

candelabra-base sockets, three bulbs, one switch, and suitable wire for wiring the lamps to the switch and to the outlet plug, together with the outlet plug

3 pieces scrap 1/4 in. plastic to hold lamp sockets in cone

PROCEDURE:
1. Saw out the base piece, saving the sawed-off corner pieces for supports for cones, as shown. Sand and polish.
2. Make paper patterns of rolled-cone pieces. It is suggested that these paper patterns be trimmed toward the narrow end of the cone to allow for a regular over-lap of the rolled pieces of about one-half inch. Saw out pieces of 1/8 in. plastic sheet according to patterns; sand and polish edges.
3. Heat sheets for cone-shaped pieces in oven at 275 degrees F. until soft enough to roll. Roll cone-shaped pieces, freehand, to size and shape already planned upon, and cool until rigid.
4. Saw off small end of cone so it meets base at desired angle. Saw out supporting triangular pieces from white scrap 1/4 in. plastic.
5. Cement triangular support pieces to base of cone piece. After cement has dried, wet-sand entire bottom area on sanding belt or flat surface so that entire bottom will touch flat against base for proper cementing.
6. Fashion three supporting pieces to be cemented inside cones about half way down from top. These pieces may be cut from any 1/4 in. scrap, and need not be precisely fitted into inside of cone, but should present sufficient surface of their edges in contact with the inner surface of the cone to form a strong bond. These pieces which should be cemented in place inside the cones, are used to support the lamp sockets, and the three 1/8 in. pistil-like rods.
7. Before cementing support pieces in place, each should have 5/8 in. hole drilled to receive the lamp socket and three 1/8 in. holes to receive the pieces of rod.
8. Cement these pieces inside cone, about midway between top and base.
9. Cement plastic "stones" to top of small rods. Then cement rods into holes provided for them in the supporting piece that has been placed inside the cone.
10. Bore holes through the base up into the cones to take the wiring for each lamp. Install wiring. Consult instructor for correct method of wiring lamps.
11. Sand flats on three 1/2 in. plastic balls, and cement to under side of base to act as "feet" for the lamp.
12. Wax and give final polish to completed piece.

NIGHT OR TV LAMP

CARVING BY:
Hood Plastics Company
Phoenix, Arizona

This lamp makes use of the light piping ability of acrylic plastics. Light comes through a slotted base, above which a carved plaque is sealed.

MATERIAL REQUIRED:
1 piece black opaque plastic 3/16 x 2-3/4 x 4-1/4 in.
2 pieces black opaque plastic 3/16 x 1-3/16 x 4 in.
2 pieces black opaque plastic 3/16 x 1-3/16 x 2-1/2 in.
4 pieces black opaque plastic 3/16 x 7/8 x 7/8 in.
1 piece clear plastic 1 x 1 x 3/8 in.
1 piece clear plastic 3/4 x 3-1/2 x 4 in.
1 cord and plug, socket, 7-watt bulb, switch, connector

PROCEDURE:
1. Cut the clear plastic block for top plaque to shape and bevel as shown in drawing.
2. Sand and polish three long edges. Wet-sand bottom edge (2-1/2 in. edge) only.
3. Cut top plate to dimension, 4 legs to size, and 4 black side strips to size.
4. Locate center of top plate, and lay out area to be slotted.

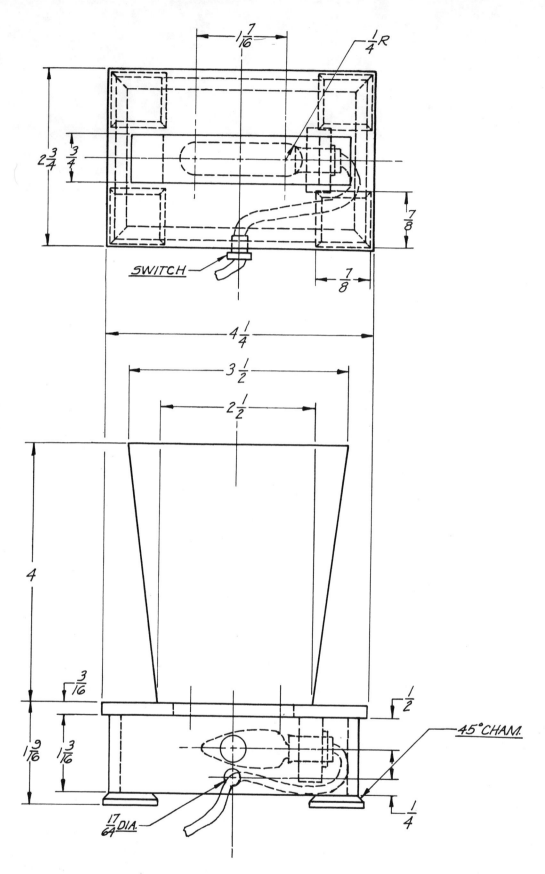

Night lamp.

Projects From Plastics

5. Drill 1/2 in. diameter holes at ends of slot.
6. Using hand coping saw or power scroll saw, cut tangent lines from hole to hole, thus forming slot. File smooth.
7. Sand and polish top plate, and bevel, sand and polish legs.
8. Miter short ends of each of the four side strips at 45 degrees, making sure they fit properly to form rectangular box.
9. Using one of the long side strips only, locate center vertical line, and locate points and drill holes for switch and cord as indicated.
10. Seal four side strips together at miters to form rectangular box. Make sure they are square at corners. Thoroughly dry.
11. After four-sided rectangular box is thoroughly dry, sand both sides of edges on disc sander so it is flat.
12. Seal top plate to four-sided rectangular box. Allow to dry.
13. Seal four legs at corners of box on bottom side.
14. Cut 3/8 in. x 1 in. x 1 in. piece for socket mount, and locate and drill 5/8 in. diameter hole through center.
15. Seal into position on inside of box, allowing proper spacing so socket and bulb when mounted, will locate bulb directly below slotted area in top plate.
16. Install switch and cord through side wall. Install socket in socket mount by merely pressing into position. Install bulb. Make wire connections.
17. Internally carve clear top plaque as desired, dye and fill.
18. Using soak method, seal plaque over slotted area. Since the area to be sealed is larger than the slot, it will seal well if properly centered. Make sure front view of carving is sealed so that switch and cord are located to back of lamp.
19. Clean and wax.

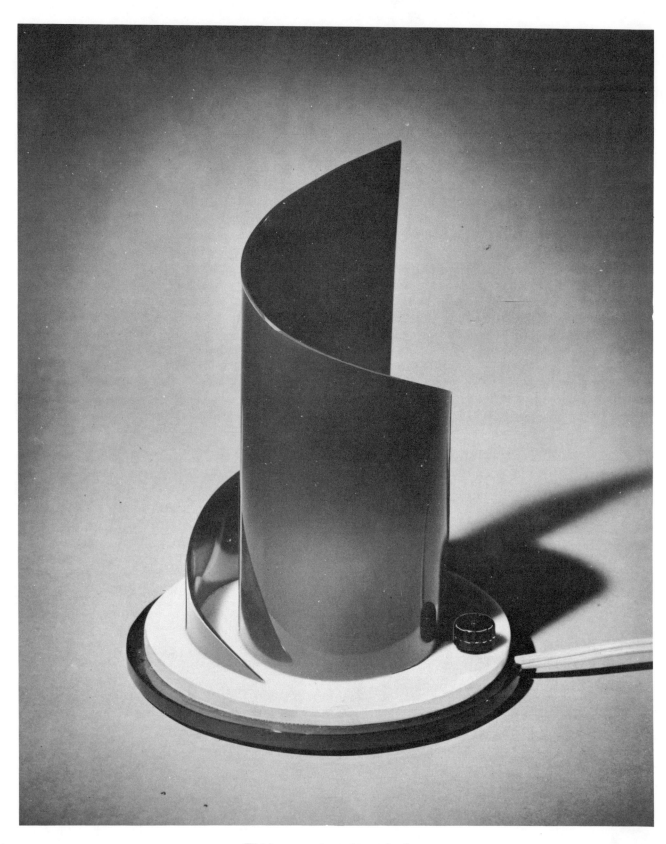

TV Lamp of modern design.

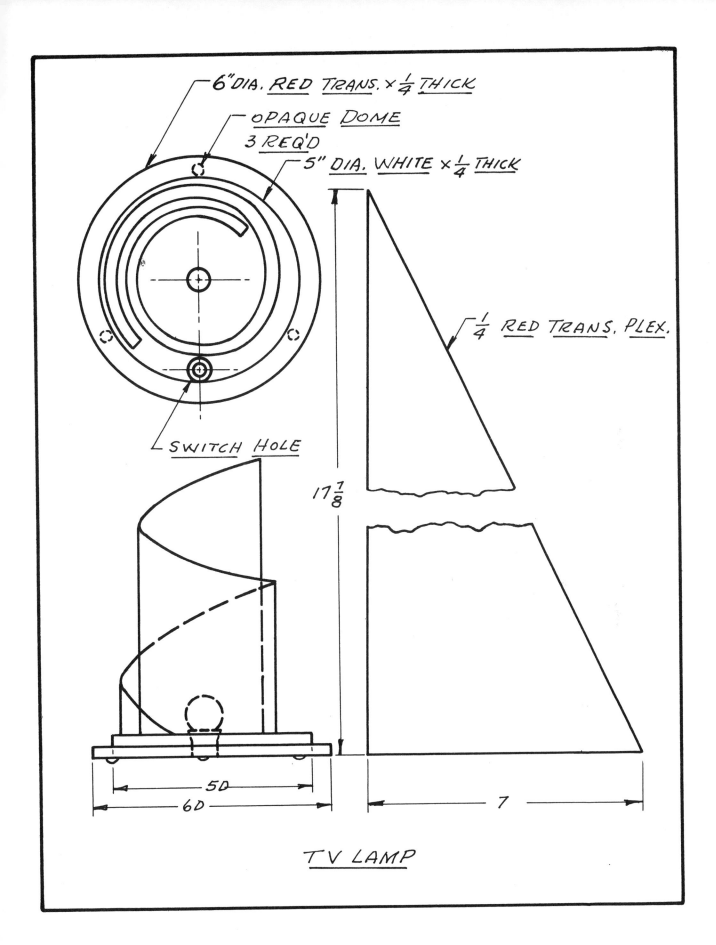

TV LAMP OF MODERN DESIGN

SUBMITTED BY:
Alice Kuhlenkamp
Arts and Crafts Instructor, High School
Port Huron, Michigan
MADE BY: Student, Gary Hudson

MATERIAL REQUIRED:
1 piece 1/16 x 7 x 17-7/8 in. Red translucent plastic (or any other suitable translucent color)
1 piece 1/4 x 6 x 6 in. Red translucent plastic
1 piece 1/4 x 5 x 5 in. White opaque plastic
Electrical parts: (1) 7-watt bulb, 1 socket for same, 1 switch, 1 - 6 ft. cord & plug
3 clear (or white opaque) domes (#61)

PROCEDURE:
1. Take piece of 1/16 in. translucent red plastic 7 in. x 17-7/8 in. and lay out diagonal line from one corner to other (this will actually be sufficient material for two such lamps).
2. Using table saw, cut diagonally across panel.
3. Using compass, draw circles on 6 in. and 5 in. panels, and cut out on band saw. Sand and polish edges.
4. Solvent seal two circles together, concentrically, after having removed all masking paper, except paper from bottom side of red base panel.
5. On bottom side of two circles sealed together, find center and drill 5/8 in. diameter hole through both pieces. Also drill a 5/8 in. diameter hole about two-thirds of way through, starting from bottom side (red side) about 1 in. in from outer edge. Using a 13/32 in. diameter drill, finish drilling this hole on through. This will give a smooth surfaced countersunk hole for inserting switch through top, and lock nut in countersunk portion.
6. 5/8 in. hole in center will allow press fit of regular 7-watt lamp socket through center.
7. Seal three domes in position on bottom side after removing this last masking paper. Domes allow cord to lie under base.
8. Make electrical connections with 6 ft. cord, switch, and socket. Check with your instructor on proper connection.
9. With masking paper removed, place 1/16 in. red panel in oven and heat for about 4 minutes at 250 degrees F. Remove while flexible, and free form this panel to desired shape, making sure that bottom rests squarely on flat surface. Allow to cool.
10. Place flat edge of formed panel in tray filled with thin skim of solvent, and allow to soak for about one minute.
11. Place in position on lamp base and allow to set.
12. Clean and wax.

MODERN FREE-FORM LAMP

SUBMITTED BY:
William Forkner
Hobbs High School
Hobbs, New Mexico

The beauty of this lamp with its thick sections of crystal clear plastic, and gleaming black base, is unsurpassed. Add to this the polished beauty of the brass center rod, and a matched shade of equal taste, and you have a modern lamp which will be admired by everyone. The decorative top can be made of colored plastics laminated together and polished after shaping, or as in the case of this lamp, the top was made up of clear strips, and sealed together with laminating dye. Then it was shaped, sanded and polished.

MATERIAL REQUIRED:
1 piece clear plastic 1-1/2 x 6 x 12 in.
1 piece black plastic 1 x 6 x 9 in.
1 polished brass tube 3/8 CD x 13 in. long, with both ends threaded and including two nuts for same

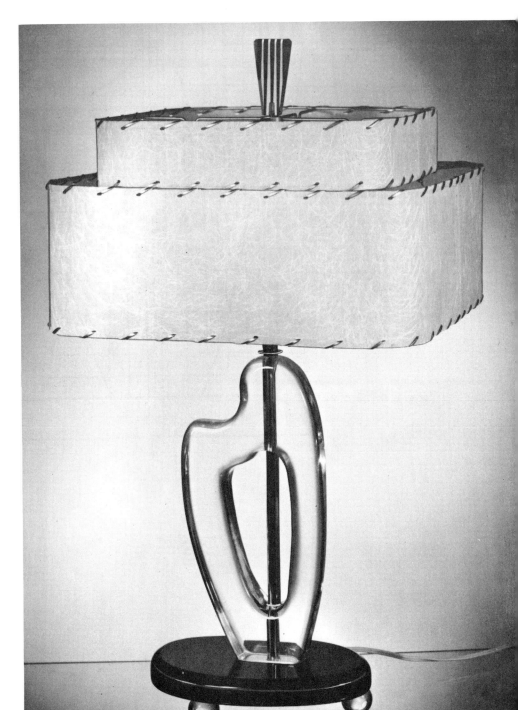

Modern free-form lamp.

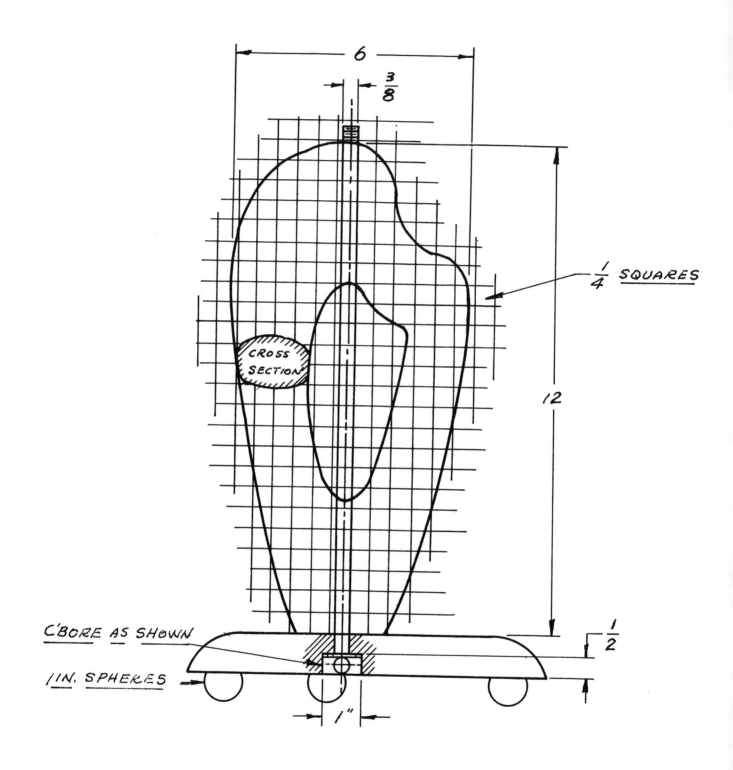

TABLE LAMP

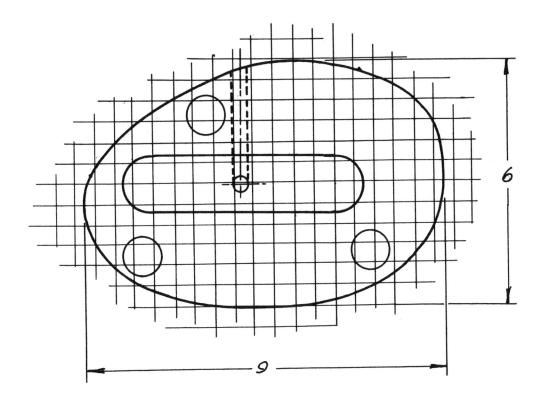

3 only 1 in. diameter clear plastic spheres
1 lamp harp (metal)
1 shade of appropriate design and color (your choice)
1 set wiring (6 ft. cord, socket, bulb, etc.)
6 pieces clear plastic 1/4 x 1-1/4 x 3 in. (for top)

PROCEDURE:
1. Make layout of lampshade design, and cut and polish edges to shape.
2. Make layout of upright section, and cut to outside shape.
3. Drill suitable hole in center section, and use scroll saw to cut out center section.
4. Sand and/or file to desired shape, and to cross section as shown. Polish fully.
5. Drill hole 13/32 in. dia. through clear section for insertion of brass tube. A length of 3/8 in. dia. steel rod may be welded to your 13/32 in. drill to extend shank.
6. Countersink hole in bottom side of black base approximately half way throught (1/2 in.), then drill 13/32 in. hole remaining distance through base. Following this drilling, drill cross hole for wiring through edge of black base as shown, taking care to see that hole comes out into countersunk portion of center hole. This cross hole should be 17/64 in. diameter.
7. Remove all masking paper.
8. Sand a flat on each of the three spheres (feet), about 1/2 in. across, and seal at suitable locations.
9. Insert brass tube through both sections and place harp in position.
10. Screw lamp socket to threaded top end.
11. Insert cord through side hole in base, then up through brass tube, and connect to lamp socket.

Preparation of top dome:
12. Using black laminating dye, seal the six pieces of 1/4 in. plastic together, and allow to dry thoroughly. Seal flats to flats, thus ending up with a block approximately 1-1/4 in. x 1-1/2 in. x 3 in. long.
13. Cut block to shape.
14. Sand and polish all surfaces.
15. Drill and tap 10-24 NC thread for assembly to lamp harp top.
16. Clean and wax all surfaces of base, upright and dome.
17. Screw in bulb of proper wattage, place shade in position, and assemble top dome.

PLASTIC ROD LAMP

SUBMITTED BY:
Eugene Steidemann
Roosevelt High School
St. Louis, Missouri

This project is unusual in that rods are used for the upright part. Because of their curved surfaces placed side by side, the entire center section is hidden from view. Even though clear rods are used, no metal center stem is visible.

A lathe is required to machine the grooved base and to shape the edges of the top and bottom plates.

MATERIAL REQUIRED:
  15 pieces clear cast plastic rod 3/8 in. dia. x 11-3/4 in. long
 *1 piece clear plastic (for top plate) 1 in. x 3-5/8 x 3-5/8 in.
 * 1 piece clear plastic (for base plate) 1-1/4 x 5-3/4 x 5-3/4 in.
  4 pieces clear plastic 1/2 x 7/8 x 1-1/4 in.
  1 piece threaded lamp pipe, 3/8 in. OD x 14-1/2 in. long with clamp nuts
  1 lamp wiring kit (socket, 6 ft. cord & plug, lamp harp)
  1 lampshade
 (*) Allows about 1/8 in. extra stock for turning in lathe to full diameter shown

PROCEDURE:
  1. Cut the 15 pieces of rod to length; square off ends.
  2. Cut top plate and base plate to square, lay out circles of proper diameters on blocks (3-1/2 in. dia. on small block and 5-5/8 in. dia. on larger block).
  3. Cut block within 1/16 in. of final diameter on band saw.
  4. Place small block (now a circular disc) in lathe, using live center pad in tail stock, and pad in chuck. Use friction hold principal to hold block in lathe, then turn outer diameter to size, and shape edge to desired grooving, or make template and turn to pattern provided in drawings. Groove face of block to a depth of 1/4 in. cutting the size as indicated (to receive rod ends).
  5. Use same method as in step 4, cut to shape, and turn the base to size. Groove as in step 4, to receive rods.
  6. Drill 25/64 in. dia. hole through center of top disc, and bottom plate to allow for insertion of threaded rod.
  7. Polish edges of top plate disc, and base plate disc.
  8. Insert threaded pipe through base plate and top plate, then insert the 3/8 in. rods in grooves in the plates. By tightening the top and base nuts in place, the 3/8 in. rods will be drawn into position, thus assembling the unit.
  9. Cut four clear feet to size, shape one short end to match lathe turned top and base plates as closely as possible. Sand and polish all 4 edges.
  10. Seal four feet onto base plate, evenly spaced. Allow to dry.
  11. Insert wiring through pipe and wire to socket in proper manner.
  12. Assemble harp into position, and tighten top nut.
  13. Clean and wax all plastic parts.
  14. Place shade in position and secure.

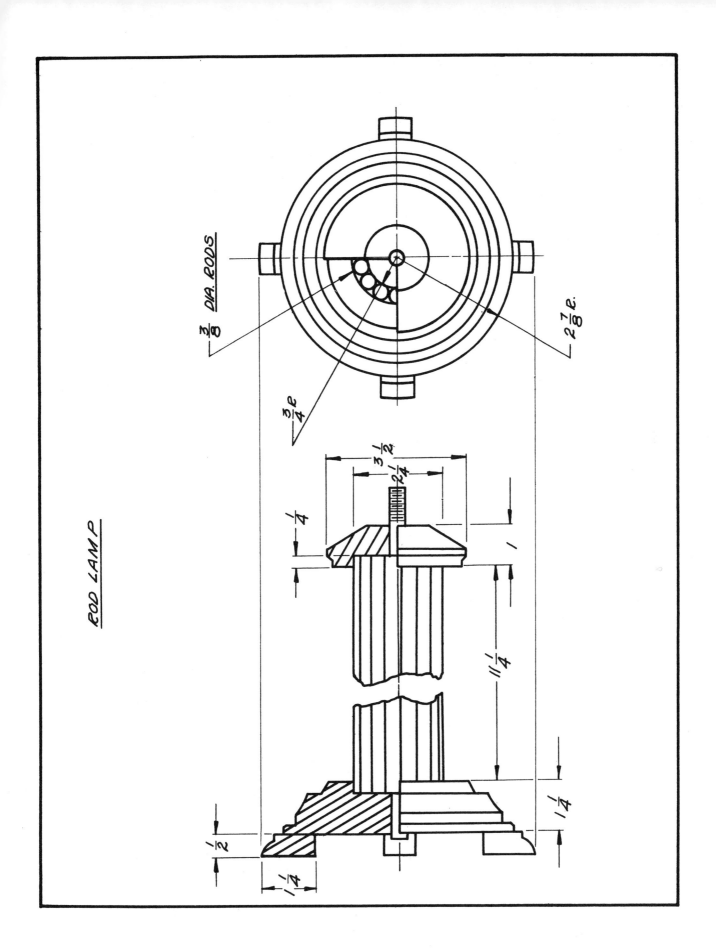

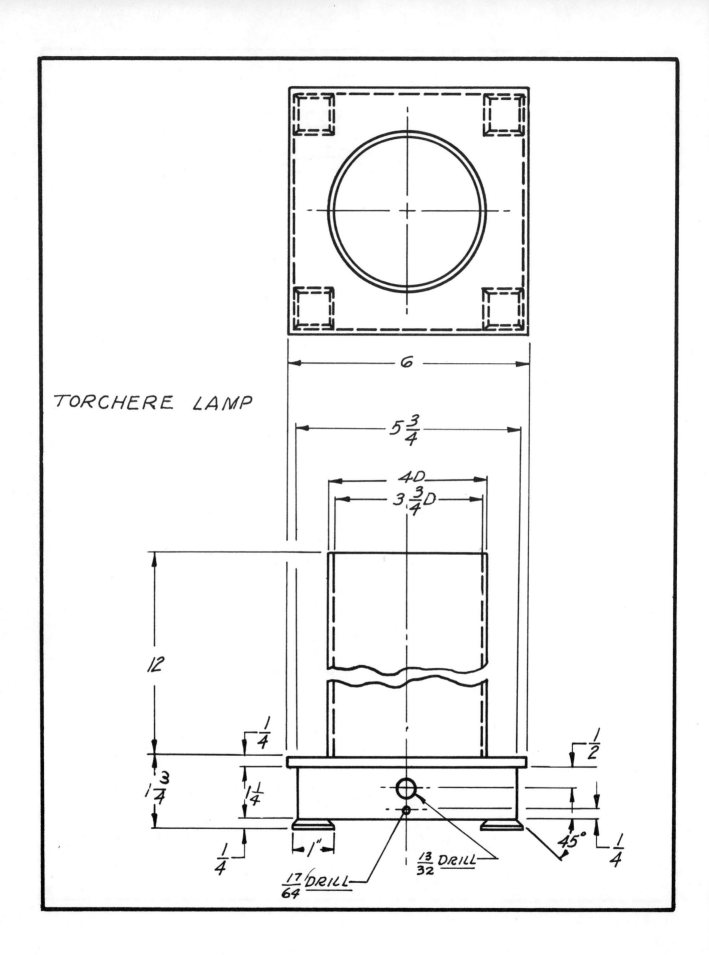

TORCHERE LAMP

Used in this lamp are three kinds of plastic, Plexiglas, Lucite and Polyplastex lamp shade material which diffuses the light. Polyplastex is available in a number of designs. Used in pairs, these lamps make attractive buffet lamps.

MATERIAL REQUIRED:
1 piece black opaque Plexiglas 1/4 x 6 x 6 in.
4 pieces black opaque Plexiglas 1/4 in. x 1-1/4 in. x 5-3/4 in.
4 pieces black opaque Plexiglas 1/4 x 1 x 1 in.
1 piece Lucite tubing 4 in. OD x 1/8 in. wall x 12 in. long
1 piece Polyplastex lamp shade material (desired pattern and color) 12 in. x 12 in.
1 porcelain socket with screw type mount
1 6 in. showcase light bulb to fit socket
1 cord, plug, switch

PROCEDURE:
1. Cut 4 in. dia. tube to length. Sand both ends flat and smooth. Polish one end only.
2. Cut 1/4 in. black Plexiglas top plate 6 in. x 6 in; cut four side strips to size; cut 1/4 in. x 1 in. x 1 in. feet to size.
3. Sand polish edges of top plate. Bevel, sand and polish four feet.
4. Locate exact center of top plate, and drill or scroll saw hole to take socket.
5. Miter ends of 1/4 in. x 5-3/4 in. side strips to 45 degrees.
6. Locate and drill 13/32 in. hole for switch installation, and 17/64 in. hole for cord insertion in one side panel only.
7. Using soak method, seal all four side plates together at miters, making sure the resultant box is square. Allow to dry.
8. When dry sand edges of box being sure parts will contact flat surface at all points.
9. Seal top plate to four-sided box. Allow to dry.
10. Install socket, switch and cord. Check with instructor on proper wiring.
11. Cement four feet at corners as indicated. Allow to dry.
12. Seal unpolished edge of tube to exact center of top plate, and allow to dry.
13. Cut desired pattern and color of Poly-

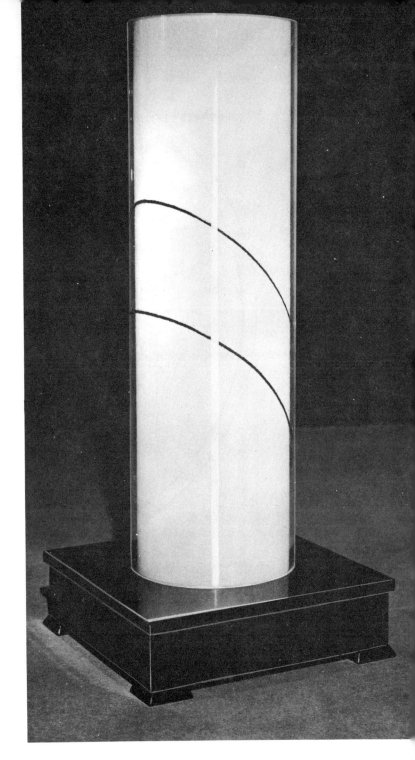

Torchere lamp.

plastex to size and insert inside of tube. Because of the slight variation of tubing diameters you may have to trim edge to a little closer dimension. However, material should "snap" exactly into place, thus holding itself into position. Do not glue.

14. Clean and wax.

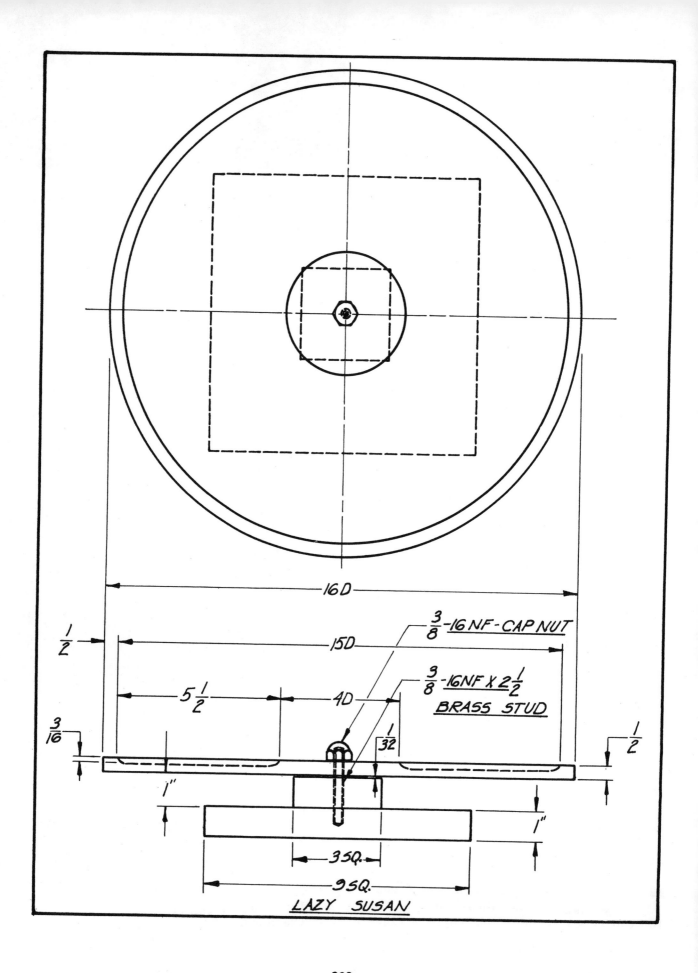

LAZY SUSAN

SUBMITTED BY:
Phil Brooks
University of Missouri
Columbia, Missouri

In this Lazy Susan of black opaque plastic a satiny finish on the center lathe turned portion forms a pleasing contrast to the other surfaces which are highly polished.

MATERIAL REQUIRED:
1 piece black plastic 1/2 x 16-1/4 x 16-1/4 in.
1 piece black plastic 1 x 3 x 3 in.
1 piece black plastic 1 x 9 x 9 in.
1 threaded brass stud 3/8 in. - 16 x 2-1/2 in. long (see note in step #13) with chrome plated decorative cap nut

PROCEDURE:
1. Cut the 3 in. square and the 9 in. square base sections to size.
2. Cut the 1/2 in. black panel to a circle roughly 16-1/16 in. diameter on band saw.
3. Sand and polish edges of two base pieces.
4. Remove masking paper from two base pieces and clean.
5. Seal 3 in. base block to exact center of 9 in. lower base section, using solvent and soak method. Allow to dry.
6. Locate center of 3 in. block and drill and tap a hole with full thread 1-1/2 in. deep. This will extend through the full thickness of the 3 in. block and 1/2 in. onto the lower base section.
7. Screw bolt down into threaded hole to full depth and tighten.
8. Place the 16 in. black disc in lathe using friction hold.
9. Use low speed, and sharp tool bit with end slightly rounded to produce smooth finish. Lathe turn outer diameter to 16 in. and break sharp edges. Lathe turn inner portion to approximately 3/16 in. depth leaving outer rim of 1/2 in. width and center raised section of 4 in. diameter. Turn and feed slowly to produce smooth satiny finish.
10. Remove from lathe, locate center and drill 25/64 in. diameter hole.
11. Remove masking paper from back side of disc.
12. Clean and wax all parts.
13. Assemble over stud protruding from base, and screw cap nut onto end of stud allowing about 1/32 in. play so tray will turn.

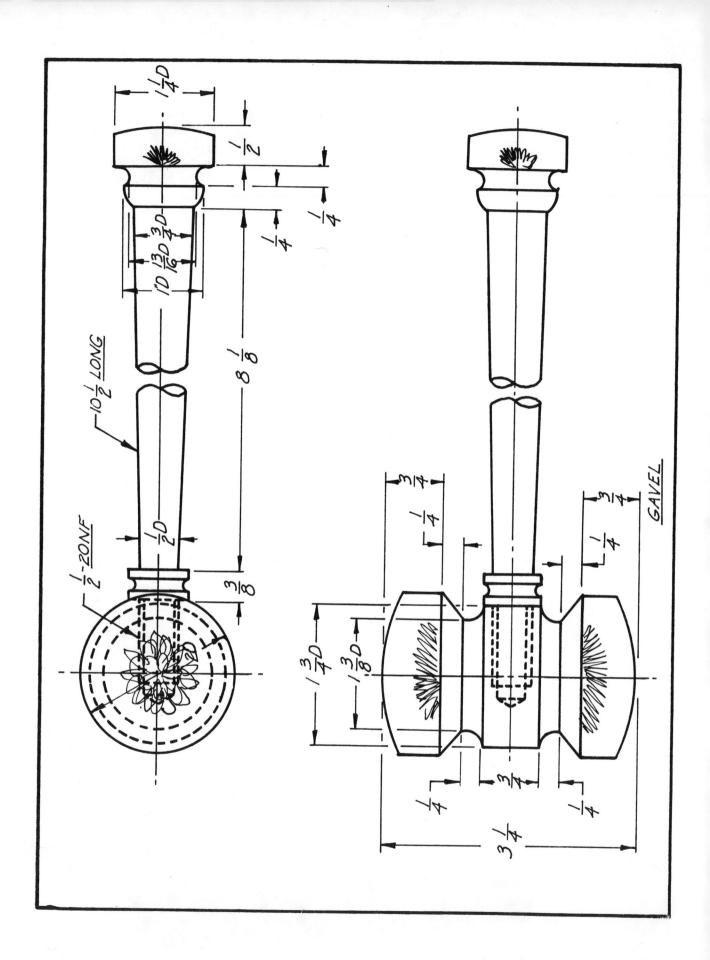

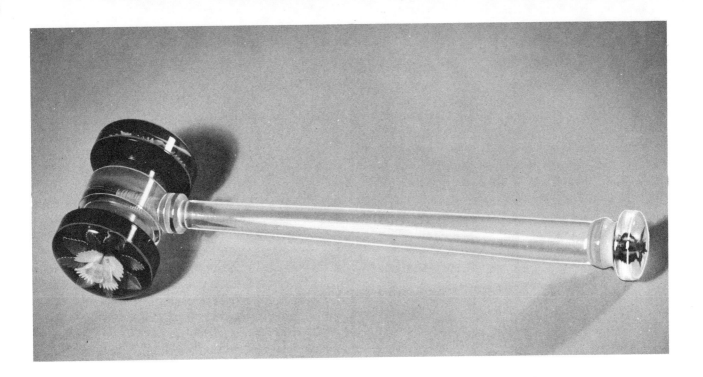

LAMINATED GAVEL

SUBMITTED BY:
Floyd Dickey
John Adams High School
South Bend, Indiana

A splendid opportunity presents itself here for the student to choose an array of color in the makeup of this highly polished gavel.

MATERIAL REQUIRED:
3 pieces clear plastic 3/4 x 2-1/8 x 2-1/8 in.
2 pieces black plastic 3/16 x 2-1/8 x 2-1/8 in.
2 pieces red fl. plastic 3/16 x 2-1/8 x 2-1/8 in.
1 piece clear plastic 1/2 x 1-3/8 x 1-3/8 in.
1 piece black plastic 1/4 x 1 x 1 in.
1 piece red fl. plastic 1/4 x 1 x 1 in.
1 piece clear cast plastic rod 3/4 in. dia. x 9-1/4 in. long

PROCEDURE:
1. Cut all pieces to sizes indicated. Remove masking papers and clean.
2. Internally carve, dye, and fill two of the 3/4 in. thick head sections and also the tip section of handle (1/2 in. x 1-3/8 in. x 1-3/8 in. block). This carving is optional, but highly decorative if used. (Sand carved surface smooth and flat after it has been filled and hardened).
3. Laminate the sections of gavel head and handle as indicated. Allow all to thoroughly dry.
4. Place laminated head section in four-jaw chuck in lathe, and turn to any lamination joint, forming curved surface on end as indicated.
5. Reverse ends, and finish lathe turning head in three-jaw chuck, meeting the previous turning at joint, thus, not showing where two cuts meet. If a slightly rounded tool bit is used, a fine satiny finish will result which can be buffed directly from this finish, requiring no sanding.
6. Place laminated handle in three-jaw chuck in lathe and turn laminated end section. Center drill opposite end. Reverse in lathe, and hold turned section in three-jaw chuck. Place opposite end in live center in tailstock, and lathe turn end to be threaded, decorative shoulder, and taper.
7. Cut 1/2 in.-20 thread on portion indicated using lathe or button die.
8. Polish all surfaces.
9. Drill and tap 1/2 in.-20 hole in head section as indicated.
10. Assemble parts, clean and wax.

CIGARETTE BOX and DISPENSER

Made of crystal clear plastic this dispenser holds almost two packs of cigarettes. When the lid is raised by pulling upward on the knob, the cigarettes are brought within easy reach. When the lid is returned to the closed position, the grooves fit down over four sides of the box tight enough to keep the cigarettes from drying out. This makes a good project for the more advanced student.

MATERIAL REQUIRED:
- 1 piece 1/8 in. clear plastic sheet, 3-3/4 in. x 12-1/2 in. long (bent to form the four sides of th box)
- 2 squares 3/16 in. plastic, 4 in. x 4 in. (bottom, lid)
- 1 rod, 3-1/2 in. long x 5/8 in. square
- 4 pieces 1/8 in. sheet, 1-1/2 in. x 2 in. (fins of dispenser)
- 1 square 1/8 in. clear plastic 2-3/4 in. x 2-3/4 in. (false bottom)
- 1 piece 3/8 in. plastic, 1-1/4 in. square (knob on lid)
- 1 piece round rod 5/16 in. x 2-5/8 in. (guide rod)
- 4 clear domes, 3/8 in. diameter, for "feet" on the box

PROCEDURE:
1. Bend long sheet of clear plastic to form four sides of box. This is best done by first cutting saw kerfs into sheet at exact line of bending. The kerfs should penetrate about one-half the thickness of the sheet. The kerfs will disappear when sheet is bent.
2. Next, heat plastic sheet, using a strip heater, with saw kerf placed directly above heating unit. When plastic is soft enough, make right-angle bend, squaring the bend carefully along some square corner so the box will have square corners when finished. Hold bend in position until cool enough to be rigid. Then proceed to next bend, until three square corners have been formed. Ends of sheet

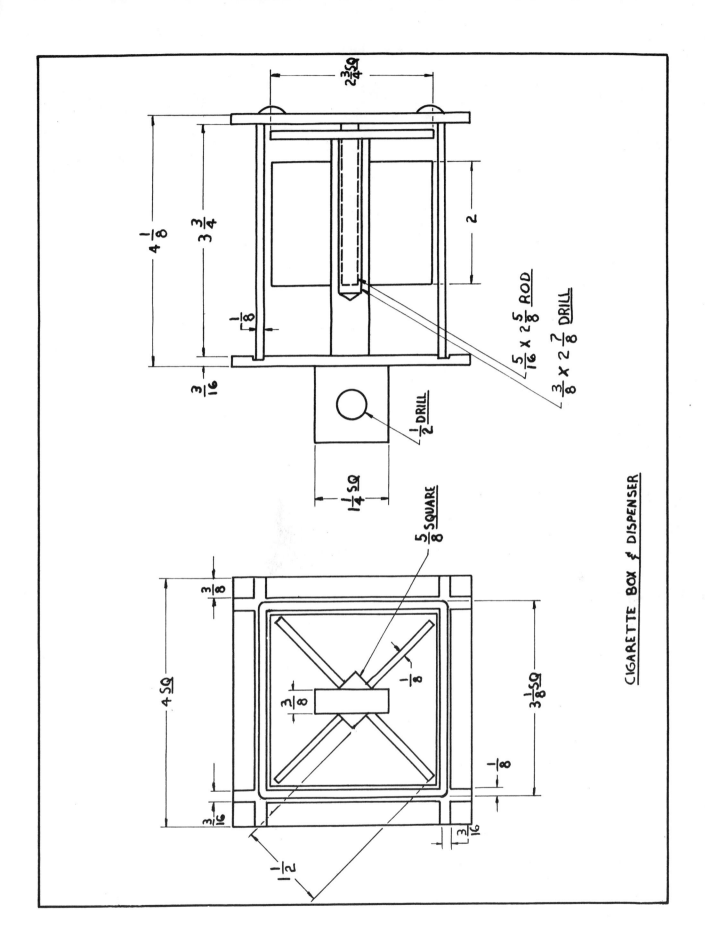

Cigarette box and dispenser.

    should then come together with the last end just lapping over first end, and butting squarely against it.
3. After making sure all bends are square, and that the butted ends, when together, will form the fourth perfect corner, cement two ends together to form four sided square tube.
4. Sand top and bottom edges of square tube.
5. Cement one rim of square tube to bottom sheet (one of the 4 in. x 4 in. squares). Carefully center the square tube to the square bottom sheet.
6. Cement four domes or feet to bottom of box.
7. Prepare lid to fit down squarely over top of box. Do this by making a series of narrow saw kerfs on under face of lid. Kerfs should be about 1/8 in. deep and 3/16 in. wide, and parallel to edges of lid. These kerf marks then act as a decoration to lid as well as provide grooves that enable lid to fit down tight over top of box.
8. Fabricate dispenser unit. See drawing and photos for details.
9. Make knob for top of dispenser from 1-1/4 in. x 3/8 in. plastic square.
10. Polish edges particularly top edges that form the top rim of the box. Wax and polish finished piece.

MUSICAL PIANO CIGARETTE DISPENSER, or JEWEL BOX

SUBMITTED BY:
Clyde Brown
Carthage High School
Carthage, Illinois

This little piano may be made of your choice of colored plastics; however, black opaque with sparkling white keys makes one of the most beautiful combinations.

MATERIAL REQUIRED:
1 - Base 1/4 x 4-1/2 x 8-3/8 in.
1 - Top 1/4 x 4-1/2 x 7-1/2 in.
1 - Straight Side Rail 1/4 x 1-1/8 x 8-1/4 in.
1 - Curved Side Rail 1/4 x 1-1/8 x 10-1/8 in.
2 - Dividers 1/4 x 1-1/8 x 3-1/2 in.
1 - Lid Support 1/4 x 1/4 x 3-1/8 in.
3 - Legs 1/4 x 1 x 3-1/16 in.
2 - Hinges 1/4 x 1/2 x 1-3/8 in.
1 - Musical Box Cover 1/8 x 3 x 3-1/4 in.
1 - Short Keys 1/16 x 1/2 x 3-3/8 in.
1 - Long Keys 1/8 x 7/8 x 3-3/8 in.
2 - Hinge Pins 1/8 x 3/8 in. (for top)
1 - Hinge Pin 1/8 x 1/4 in. (for prop)
1 - Swiss Musical Unit

PROCEDURE:
Because of the numerous parts required for this project we have felt it advisable to label the parts with letters for ease of identification.

1. Make a complete set of patterns.
2. Lay out patterns on plastic, and cut parts to shape.
3. Bevel top section A on all edges.
4. Sand and polish all edges of all parts except the following which are to be left sanded only:
   Long bottom edge of rail C.
   Long bottom edge of rail D.
   Flat edge of two hinges I.
   Two ends and one long edge of keyboard material, J and K.
   All edges of part L.
   Top flat edges of legs H.
   Two ends and bottom straight edge of E.
   Two ends and one long edge of divider G.
   All other edges of all parts are to be polished.
5. Remove masking paper and clean parts.
6. Place part D in oven at about 275 degrees F. and heat and bend to shape as drawn by outline M. Allow to cool in position.
7. Cement rail piece C to base B 3/8 in. in from long straight edge.
8. Cement cross member G to base and to rail C at point 1-5/8 in. back from front straight edge.
9. Cement curved side rail D to base, extending around curve and capping over squared end of straight rail, C.
10. Drill hole in cross member E and prop F, and install hinge pin permanently in E.
11. Cement cross member E into position, parallel to G, and allowing 2-3/4 in. between two members.
12. Cement legs H to base in positions indicated.
13. Locate and drill holes in hinges I and matching holes in piece D on capped end and on notched end near front.
14. Insert pins and cement hinges to lid A.
15. Install Swiss musical unit in curved recess area in base, drilling hole where necessary in base for winding handle to come through on bottom side.
16. Make automatic on-off lever from paper clip and secure to musical unit shut-off mechanism.

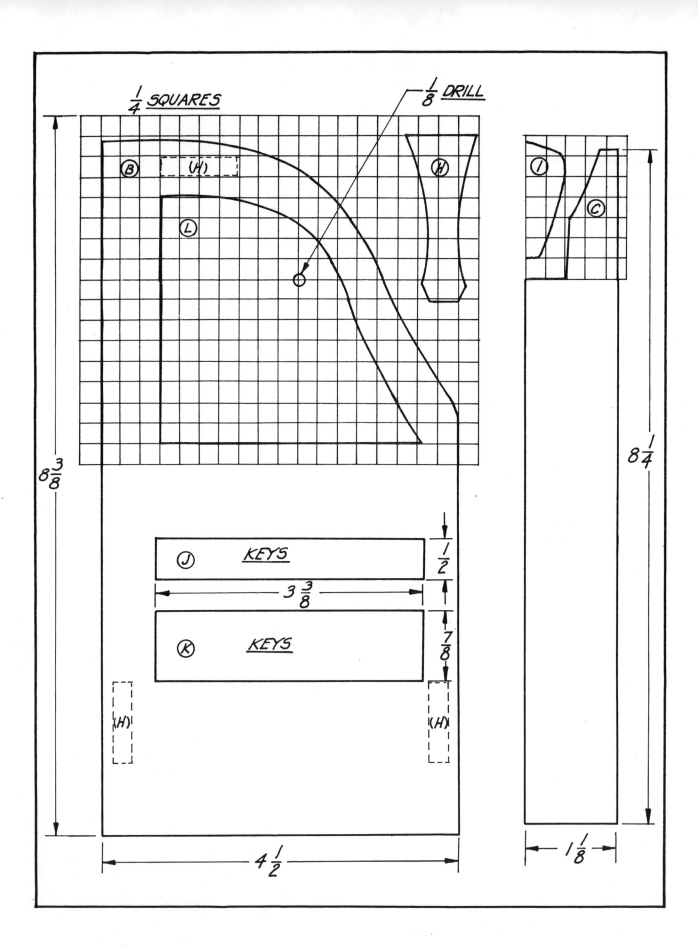

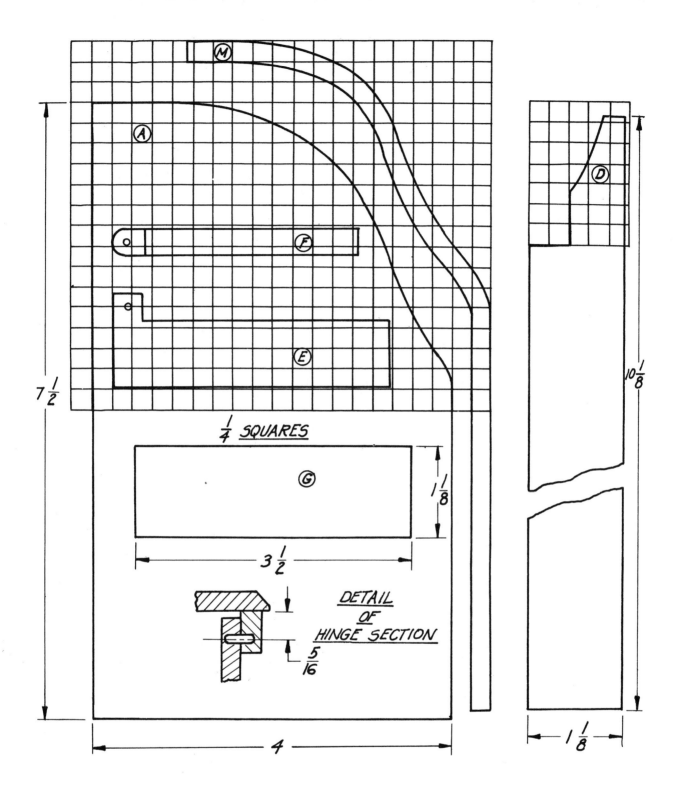

17. Locate and drill hole in panel L directly over the on-off lever of musical unit.
18. Cement piece L to piece D (rail) down into recess, allowing about 3/16 in. space to top edge of rail.
19. To prepare the keyboard unit, cement piece J (black) to piece K (white) so that two long edges match. Allow to dry thoroughly.
20. Make saw kerfs into this cemented unit J-K, 3/16 in. apart and slightly into depth of white, through the black (approximately 1/8 in.).
21. Cement keyboard to base between rails as indicated.
22. Apply coat of wax and polish.

229

INTERNAL CARVING AND COLORING OF ACRYLICS

Because of the extremely good transparency of acrylic plastic, such as Plexiglas, it is possible to carve out a figure on the inside of a block of the material, and then view the carved figure in three dimensions through the clear plastic.

In this internal carving, the figure is really a hollowed-out place in the clear material--it is like sculpturing, only in reverse.

Internal carving of plastics offers such unusual opportunities for creating colorful objects of great beauty, that the subject will be given thorough coverage, in this book.

BASIC REQUIREMENTS

In doing internal carving of plastics you will need:
1. An electrically driven, high speed rotary cutting tool, such as the Foredom Flexible Shaft Machine, Fig. 36, or, the Handee Motor Tool, Fig. 37.
2. Long tapered internal carving drill, metal burrs, Fig. 38.
3. Quantity of thick, clear, salvage plastic for practice carving (type that may be carved and dyed), Fig. 39.
4. Jars of the more popular colors of dyes.
5. Plaster and blending granules for filling cavities.
6. Solvent cements for fastening backings and metal findings to finished carvings.

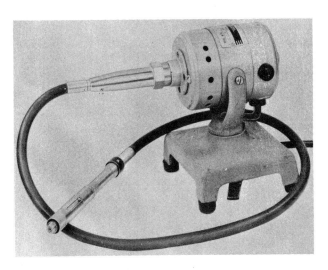

Fig. 36. Foredom flexible shaft machine for internal carving of plastics.

Fig. 37. The Handee motor tool, a small direct-drive tool for internal carving.

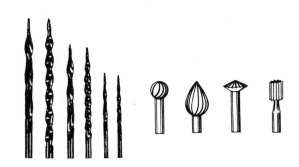

Fig. 38. Left. Tapered internal carving drills. Right. Steel burrs.

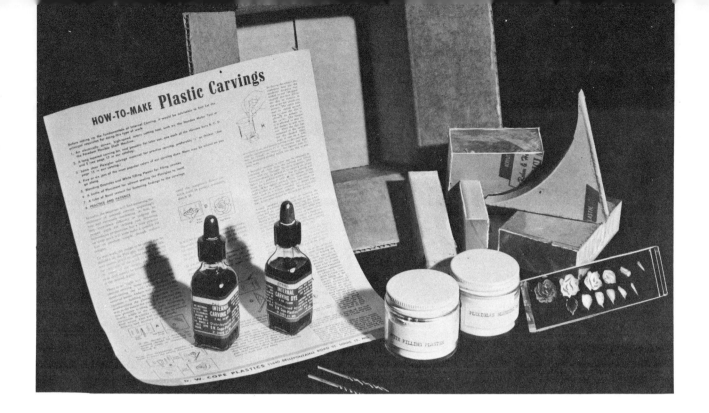

Fig. 39. Internal carving kit for beginners, which includes progressive carving block, scrap Plexiglas for practice carving, dyes, carving drill, blending granules, filling plaster, instruction sheet.

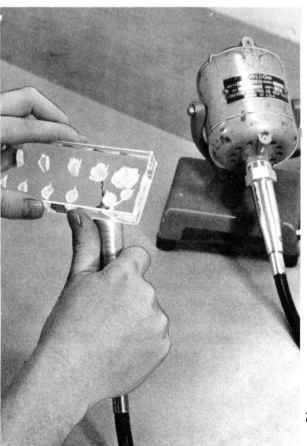

Fig. 40. Proper positioning of the hand for the beginner in internal carving. Note thumb in position serving as control factor as well as safety factor.

7. A willingness to exercise patience, and do a lot of practicing.

FIRST STEP

One of the first things the beginner at internal carving must do is to learn the _feel_ of the high-speed tool as it bites into plastic material. The cutting tools, when used in the machines recommended turn at the rate of 12,000 to 18,000 r.p.m. A cutting tool--even a small one--rotating at such high speed can be dangerous. It can cut rapidly into plastic-- it can cut faster into human flesh, and even bone--so watch it!

The cutting tool most used for carving is the sharp-pointed drill, It is the most versatile of all the tools and is used for all deep cuts in internal carving, such as most flower petals, leaves, etc. But it is well to become familiar with some of the other tools as well. Experimenting with these will help to give the "feel" of working with plastic, without the obvious danger of using the sharp, deep-cutting pointed drill. The "feel" is the most important part of learning to use tools in plastic carving.

It is well to start with a small ball burr. Insert the shaft of the burr in the chuck of your cutting tool. Take a scrap of acrylic plastic (you may use one of the thinner pieces for this work). Begin manipulating the cutting tool in contact with the plastic. Your first sensation will be that the tool is difficult to control. At the high rotary speed of the cutting tool, the burr seems to act erratically, trying to dash off in various directions. To overcome this tendency, learn to control the relative motion of the tool and the plastic being carved. You must brace the hand holding the cutting tool against the plastic piece. Hold the tool and the workpiece in a constant and planned relationship with each other.

There is no one particular right way to achieve control of the cutting tool. Some workers prefer to clamp the workpiece rigidly to the workbench and rest the hand holding the tool against the bench, meanwhile holding the tool somewhat as you would a pencil. Others like to clamp the tool in a rigid position and move the workpiece into the tool, meanwhile bracing the hand holding the work against the bench or the vise, or some other steady object.

Most expert carvers prefer to hold the workpiece in one hand (the left, if you are right-handed) and the tool in the other so that both the tool and the workpiece can be manipulated. See Fig. 40. This gives the carver a great deal of flexibility of movement. With this method of handling, most carvers brace the hand holding the cutting tool by resting the thumb of that hand against the edge of the workpiece--much as one would do when peeling a potato. If you are carving in the center of a large piece, you can brace the thumb against the hand holding the workpiece, or anchor it against the edge of a design already cut. Some workers fasten a small rubber suction cup against the under side of the piece as an anchor point.

MUST CONTROL CUTTING TOOL

No matter what method of bracing you use, you must have absolute control of the travel of the cutting tool, and be able to start or stop a cut at any desired point, or at any instant. The necessity of being able to control your cutting tool absolutely and at all times, cannot be overemphasized.

With the small ball burr in your cutting tool, continue practice cutting on a piece of scrap plastic. The burr is not intended for actual internal carving, but practicing with it will give you the feel of the cutting tool as it bites into the plastic. Try carving letters, numerals, and the like on the under side of the piece, while watching your work through the top. Guide the cutting point by manipulating the drill, all the while observing the precaution of keeping the thumb of the hand holding the drill well braced against the side of the workpiece. Best control is obtained if you draw the cutting tool toward the thumb with a squeezing motion of the hand. Again we liken the action to peeling a potato. In this way you have the most complete control of the cutting point. You will notice a tendency for the cutting point to wander or skid off to one side or the other. This tendency must be carefully controlled by the hand holding the cutting tool.

Practice in this manner until you have learned the feel of the different forces that try to direct the cutting tool as you make the cut. Be careful, to learn to judge the depth of the cut--depth must be controlled as accurately as the actual cutting stroke, since the carved image will be in three dimensions, and depth is one of them. Cutting with the burr will not tend to pierce through the workpiece, but with the sharp-pointed drill you can zip up through a piece of plastic with surprising speed. Learn to watch depth of cut from the beginning of your work--it will pay big dividends later.

Practice surface carving, holding the grinder as you would a pen, Fig. 41.

We recommend several hours practice with the simple burrs such as the ball burr, the wheel burr, the pear-shaped burr, and the end-mill burr. Learn how each of these points tends to cut into the plastic--how it tries to "get way from you" as you make your cut--and how to compensate for these forces by the way you hold and guide the tool.

As soon as you are certain you have mastered the general "feel" of cutting in plastic with a rotating tool, and know that you have

Internal Carving and Coloring of Acrylics

complete control of the cutting point, you are ready to experiment with the sharp-pointed drill--the real tool of internal carving.

If you examine this drill carefully, you will see that it is sharp-pointed, with grooves spiraling from the point toward the shank. It looks much like an ordinary twist drill, except for the tapered shape. A satisfactory carving drill can be made by grinding a twist drill of the proper size to a point--then sharpening the leading edges of the flutes (the ridges that spiral along the drill) with a small abrasive wheel. In sharpening the flutes re-

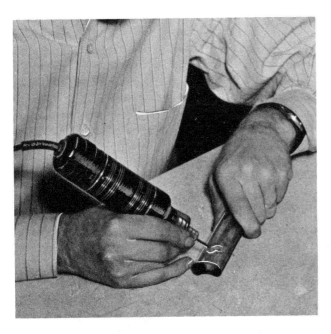

Fig. 41. Surface carving as well as internal carving can be done in this manner, using tool as you would a pen.

member the basic thing about sharpening tools for working with plastic. The cutting edge must be approximately at right angles with the surface of the plastic so that a scraping rather than an actual cutting action takes place. This is also true for the flutes of the carving drill. If you make a carving drill from an ordinary drill, most of the flutes are removed, and these must be ground back onto the drill by cutting the grooves deeper with a small grinding wheel. When making a cut sideways with the drill, the flutes do all the actual cutting.

SPECIAL DRILLS RECOMMENDED

In general it is recommended that the beginner use one of the drill bits specially made for internal carving of plastics, These drills have better and faster cutting action, and will do a neater job with less bother. It is a good idea to know the fundamentals of grinding a drill properly so that you can prepare your own drill points in an emergency, or can sharpen a broken drill point to make it useful for further work. A study of Fig. 13, page 39, will help you to master the art of sharpening a drill properly.

We mentioned the subject of broken drill points. This is one of the bothersome angles of internal carving. Drills are likely to break if you apply too much side pressure, or if they penetrate through the workpiece, or if you break through from one cut into another. Sometimes a drill so broken may be resharpened by grinding it to a new point. More often they break off too short for resharpening. Since these drills are rather expensive, every effort should be made to avoid breaking them. When you break a drill you may lose control of your tool temporarily and you may jab the end of the broken drill into your hand or wrist. Be careful!

Another thing to remember with drills is that they can become dull during use. A dull drill will not cut fast enough, and the friction of the rotating tool tends to soften the plastic in the vicinity of the cut, making the cut erratic and irregular. If the plastic softens too much it may flow in and harden around the drill point causing it to "seize." When this happens you may break the drill, or the flexible shaft, or stall the motor till it burns out, or (at the mildest) ruin the work with irregular and unsightly cuts. For this reason, every student and hobbyist should have a small, motor-driven abrasive wheel and learn to sharpen dull bits, if he intends to do much carving.

We are now ready to practice actual carving in a piece of plastic with a carving drill. Do your first practicing with a fairly thick piece of plastic--one thick enough that the drill will not go entirely through. Select a drill 1/8 in., 3/32 in. or 1/16 in. in diameter. The smaller drill is used for the smaller and more intricate cuts in advanced carving work; the largest drill is rather cumbersome to handle and cuts slowly. We recommend that the beginner start with the 3/32 in. drill as this is the best drill for practice work.

PLASTIC SHOULD BE POLISHED

Since internal carving is truly three-dimensional work, the carver must know how deeply his drill is penetrating into the plastic. Since this is difficult to judge when looking down through the work (especially for the beginner) it is a good idea to take time, before starting to carve, to polish all sides of the practice piece. In this way the carver can view his work from all sides and judge the depth of the cuts in every direction.

CARVING ROSES

One of the most attractive carved objects, yet one of the easiest to master, is the rose. We will describe rose-carving procedure step by step. By the time you are ready for this exercise you should have mastered the art of holding the workpiece and moving the drill so you can manipulate the drill point in such a way its cutting action is made as though the drill point were pivoted at the place where the tool enters the lower face of the plastic workpiece. This pivoting action is necessary to keep from unduly enlarging the hole in the bottom of the plastic--remember that the flutes cut along the entire length of the drill. Figs. 42, 43, 44.

For the first cut in the rose design, insert the drill into the plastic from the bottom side, as shown in Fig. 42. Look through the plastic workpiece from the side to determine the depth to which the drill is inserted. Most beginners tend to push the drill too far into the plastic. This will make the center of the rose stand up too far when viewed from one side. Rotate the point of the drill as shown in Figs. 42 and 43. The cut you make here is almost cone-shaped, with the base of the cone being at the top, and the point or apex of the cone being at the point where the drill enters

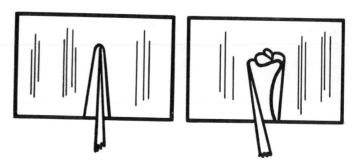

Fig. 42. Starting the rose bud cut and the primary petals.

Fig. 43. Continuing the "pivot" cut and successive layers of petals.

Fig. 44. General shaping and carving of the leaf.

Internal Carving and Coloring of Acrylics

Fig. 45. Progressive steps in carving the rose, the leaf, and the combination.

the workpiece. Practicing this initial cut a number of times will help give you the proper wrist action in manipulating both the drill and the workpiece so that you learn to "pivot" the tool at the lower opening into the plastic.

Next cuts are the surrounding petals. These are started by inserting the drill into the body of the plastic at a somewhat shallower angle to the base than used in the first cut. The petal cuts should be made smoothly, with a fan-like sweep of the drill tip. The petals are cupped about the first or center cut, with a slight amount of curling or curving to simulate the waviness of the natural rose petal. Care should be taken in making each petal not to cut into the opening made by any previous cut. A realistic effect is produced if the leaf of the petal is slightly curved in toward the center during the last part of the cut. This also enables the operator to start a new cut in behind this curl, without having to widen out the row of petals too much to prevent cutting into the preceding row. Each petal cut is started a bit behind and about one-third of the way from the final end of the preceding cut.

Three of four petals are cut about the original center cut--then a new row is started behind these, with the drill entering the plastic at an even shallower angle to the bottom face. This process is continued until the final outer row of petals is made.

Fig. 46. Depth carving in solid Plexiglas block. Because of the many reflections a single rose becomes a "bouquet" when the block is set an at angle.

235

It is suggested that before trying to carve a complete rose, the beginner practice on the sweeping, curved fan cuts at ever lower angles, to acquire the feel of making such cuts and learn the proper hand and wrist motions needed. Having mastered these, the experimenter is ready to try carving a complete rose. Practice will soon perfect your technique to the point where you can carve a very life-like rose, and you will find the results enough reward for all the practice required to reach this state of perfection. See Figures 44, 45, 46.

Most carvers of roses wish to add a stem and leaves--perhaps even to the point of showing a few thorns on the rose stem. The stem is relatively easy. You may use the carving drill, or you may find it easier to make this cut with the small ball burr or the tip of the cone-shaped burr. In any event, the stem is merely a line carved on the under surface of the plastic, beginning about 1/16 in. from the outer edge of the last petal cut. (This matter of leaving a small space between the last of the petal cuts and the beginning of the stem cut is important when it comes to dyeing the carving because if the cuts were allowed to come together, the dye would travel from the rose to the stem.

Thorns may be simulated on the stem by a single short thrust of the sharp point of the carving drill from the stem cut up into the body of the plastic for a fraction of an inch. This is a refinement you will want to add to later carvings, but is not of great importance in your first trials. The typical, saw-edge or serrated leaves of the rose are easily carved, once you have mastered the feel of the carving drill and acquired a sense of the depth to which the drill is penetrating. To carve a leaf, start the drill about 1/16 in. to 1/8 in. out from the edge of the stem cut. Thrust the drill into the plastic at a suitable angle and to a depth equal to the full length of the leaf. This cut now represents the center line of the leaf. Withdraw the drill slightly, then pull slightly toward you, or rather toward the bracing thumb on the edge of the plastic workpiece, then thrust the drill into the plastic about half as far as you withdrew it. In this manner make a succession of cuts, each slightly less deep then the preceding cut. At the same time cup the near edge of the leaf up toward the top or near face of the workpiece. The final cut on this one half of the leaf should end as the drill enters the stem cut. Now, thrust the drill back into the first cut and by cutting away from this center cut in the opposite direction complete the other half of the leaf-- again finishing the last cut in the stem-line of the figure. Procede in like manner with any other leaf cuts desired. A very lifelike reproduction of the serrated rose leaf can be made by using the technique outlined here.

Next, it is suggested that the experimenter try carving a rose at three-quarter view. This, and other carving projects may best be done by sketching the design on the top surface of the plastic workpiece with pen and ink, wax crayon, pencil, or carbon paper. Tricks of perspective--that is the illusion of distance in the carving--must be learned and applied to this work. Plan each stroke of the carving in advance carefully.

CARVING OTHER DESIGNS.

Another flower design that makes a striking piece of art work is the orchid. This design offers a great opportunity for color work in tinting the final flower. The advanced student in carving may wish to try various kinds of fish, flying birds, wild ducks, butterflies, and other designs. Details of this work may be obtained from books devoted expressly to internal carving--or the experimenter may work out his own designs with practice and patience. See Fig. 47.

As the student has observed, the action of the high-speed tools used in carving acrylics leaves the carved portion with a white, frosted appearance. This frosty whiteness creates startling and pleasing effects. Some carvers prefer to leave their carved designs set off by this white translucence. In fact, many carved designs reach their highest perfection if left in the white, uncolored state. Many carvers prefer to add the further striking effect of color to their designs. This is particularly appealing in the various flower designs, when the blossoms represented may be presented not only in three dimensions, but

Internal Carving and Coloring of Acrylics

in natural colors. The same applies to many of the other designs such as carved fish, birds and butterflies, which lend themselves well to color treatment.

USING DYES

While it is possible for the student or home craftsman to prepare dyes for carving acrylics, there are so many excellent dyes now on the market, made for this very purpose that the beginner is advised to stay with these approved products. Such small amounts of dyes are used that their cost is insignificant when you consider the superior results you get from using these special acrylic dyes.

Acrylic dyes act by penetrating the surface of the acrylic at the point of application very slightly, due to a solvent action. For this reason, the dyes work best on a roughened surface, such as the frosty interior surface of a carving. If the dye gets on a polished or buffed surface, it may be buffed off readily, or if wiped off immediately, will not stain the piece.

Some carvers prefer to remove all chips from the interior of the carving before dyeing. When this is done, the colored carving has a translucence that is very pleasing for certain effects. On the other hand, if the chips are left in, they too become dyed, giving the colored object a more solid, and in some instances a more true-to-life look.

In either case the dye is applied to the cavity with a medicine dropper, or hypodermic syringe, and spreads to all parts of the cavity by capillary attraction. If the chips are left in more dye will be required, since the chips also soak up dye.

The intensity or darkness of the color may be varied by varying the time the dye is in contact with the carved surface. The longer the dye remains in the cavity, the darker is the color. With most of the dyes supplied for internal carving colors, the dye should be allowed to remain in the carving for about one minute. After that time the remainder of the liquid dye is poured out, preferably onto a soft absorbent cloth, and the dyed cavity is allowed to dry for fifteen to thirty minutes before proceeding to fill the carved cavity with filler.

Fig. 47. Powder box imported from France, showing an interesting example of carving in plastic.

BLENDING GRANULES

Another variation of coloring may be obtained by clearing the cavity of all chips and then refilling with Plexiglas Blending Granules. To clean out all chips, the cavity is turned upside down and tapped on a padded surface. If necessary, further cleaning may be done with a sharp-pointed instrument, or by running the carving drill rapidly around through the previous cuts, being careful not to do any additional cutting. The blending granules, which are actually very tiny balls or spheres of Plexiglas, are then packed into the cavity. This is done by pouring the granules into the cavity, and then tapping the workpiece, with opening up, on a padded surface to pack the granules in place.

Because the blending granules take up much of the dye, it is possible with a filled cavity to do what might be described as differential or varied dyeing. For example, suppose we have a carved rose. If some yellow dye is applied to the center, the dye flows down through the granules to the center of the rose, and by capillary attraction spreads outward, slightly, to the granules on each side giving a gradually fading color to the granules on each side of the up-and-down line where the dye was first applied. If red dye is now filled into the remainder of the granule-filled cavity, the red travels to the parts not dyed, and at the same time blends into the faint yellow parts that have already been partially dyed by the yellow color. The overall effect is a rose with yellow center, slightly tinged with red, with the yellow gradually blending with the red, until at the outer edges only the red is seen. Jobs tinted in this fashion should be allowed to dry for a full hour or more.

By using this blending granule technique two or more colors may be blended in a given carving. With patience and practice the experimenter may achieve some very pleasing effects. Here is another case where practice is required to get the desired results.

Another color effect possible with straight dyes alone, is made by diluting all or part of the dye with denatured (ethyl) alcohol. (Do not use methyl, the radiator anti-freeze type, alcohol). This dilution lightens the color of the dye and enables the experimenter to get pastel shades of any desired intensity.

To get two or more colors in the same cavity, such as in coloring a landscape, another method is to tint the darker color, then recut the lighter portion and color it the second shade. Or, you may carve only that portion that is to be tinted one color--color it--and then carve the next colored portion and tint it. To make the tips of flower petals one color, and the remainder another color, first fill the cavity with the tip color--then pour the dye out and immediately apply the second color. Capillary attraction will hold the first color in the tips regardless of the second color added.

It is usually desirable to fill the carved cavity after tinting except, as noted previously, where extreme translucency of the tinted design is wanted. This is usually done by adding white filling plaster to the cavity. There are two ways of doing this. One is to heap the dry filling powder into the cavity while the dye is wet. The powder absorbs quite a bit of the color and gives a beautiful powdery appearance to the figure. After allowing this powder and dye mixture to stand for awhile, a small amount of water is sprinkled on the outer part of the filling to assure a hard "shell" of the dried plaster surface.

In a second method of filling, the filling plaster is mixed to a paste with water and flowed or pressed into the cavity after the dye has dried thoroughly.

In any case, the dried plaster filling should be sanded or smoothed off until the back surface is perfectly flat. In many instances it is desirable to cement an opaque piece to the back of the carved piece. This protects the plaster in the cavity filling and gives the piece a rich look that enhances its appearance.

CARVING THIN STOCK

After you have mastered the art of internal carving in thick pieces of Plexiglas, you may find it interesting to do miniature carvings in blocks of thinner material--such as 1/4 in. stock. This must be done with smaller drills, using great care not to drill entirely through the sheet of thinner material. Carving in this thinner material is appropriate for making pieces of costume jewelry.

Many attractive designs may be made by surface carving on the back of sheet material. Most such designs are not true "internal carving" since they are merely cut in the reverse surface of a relatively thin sheet of plastic material. Such designs lend themselves beautifully to what is known in the trade as "spot colorings." In this work the engraved or carved design is cleared of all chips. Then the color is applied to the design. Most artists using this sort of medium prefer to work by viewing their design through the front, or uncarved side. Any good enamel type of painting material is suitable for this sort of work. Lacquers, because of their thin consistency and the solvent action of the lacquer thinner, are not suitable for spot coloring, as a general rule. They tend to run because of capillary attraction between the solvent and the carved plastic surface. Spot-colored surface carvings, engraved and colored on the back side of the piece of clear plastic lend themselves well to the production of costume jewelry with a colorful motif. Such pieces when done on thicker plastic base stock also lend themselves well to edge lighting, since the spot-colored areas, covered with good grade enamels, show up beautifully under the glow of the light piped up from the edge of the Plexiglas block.

JEWELRY FINDINGS

Most costume jewelry calls for the fastening of "findings" to the back surface of the plastic piece. "Findings" is a term used to designate small parts usually of metal, that are to be attached to the fabricated pieces. They include such things as metal pin backs, which are attached to the back of a carved or otherwise decorated piece to make it usable as a lapel pin or brooch--metal pieces that make earrings out of plastic blocks and to metal clips which make tie clasps or holders out of decorated pieces.

These metal parts may be readily bonded to acrylic plastic pieces by using the cement referred to at the beginning of this chapter. This cement is similar to that sold for making model airplanes and various household cements. It usually consists of a cellulose nitrate-acetate material dissolved in a volatile solvent, such as methyl-ethyl ketone or acetone. With a cement of this type metal findings may be bonded to the back of a carved piece of plastic--either with or without a backing piece. Some findings for acrylics pieces are so made that small metal tangs, sticking out from the finding itself, may be heated and forced into the plastic without using a cement.

A thin, opaque backing sheet to the carving will hide the metal finding when the piece is viewed from the front. In some instances the carving, itself, will serve to hide the attached finding without using a backing piece.

LAMINATING DYES

Another interesting phase of the dyeing or coloring of plastics involves the use of the so-called laminating dyes. These materials are solvent-cement liquids to which a small amount of coloring dye has been added. If this dye is of the fluorescent type, particularly startling results may be obtained.

In use, the colored laminating cement is used to cement two (or more) pieces of clear plastic together. Care must be taken to work out all air bubbles, and the cement must be allowed to harden completely before any further work is done. After cementing, the color shows through if you look at the clear plastic straight-on. If the laminate is viewed from the edge, parallel to the cemented faces, it appears clear, with only a thin line of color denoting the line of gluing. If the laminate is now rounded off on the edges, the high refractive angle of the plastic causes the color to show through, no matter what angle the material is viewed from, and the piece appears to have the color of the laminating dye all the way through it. For example, if two pieces of plastic three inches square by one-half inch thick are cemented with a laminating dye, and after the dye has dried completely the cemented piece is sawed out into a heart shape and rounded off on all edges, the entire piece takes on the color of the dyed layer. If a red dye has been used, the whole heart-

Fig. 48. External carving or sculpturing in Plexiglas results in some beautiful carvings. Big Red was done by Horace Moore, Philadelphia, Pa; the three dogs by Lloyd Schmidt, Wichita, Kans.

shaped piece appears to be a beautiful translucent red.

Using this technique some very startling color effects can be obtained. It would be rewarding to the experimenter in plasticraft to experiment with these dyes in trying to achieve unusual color effects.

Other colored combinations are possible by combining various colored plastic parts. Some of these color combinations have already been treated in the chapter on acrylic projects. Acrylics may be obtained in various colors, in transparent, translucent, and opaque, and also in several daylight fluorescent colors which are made by adding fluorescent dyes to the clear plastic during manufacture.

PAINTING ACRYLICS

Clear acrylic also lends itself remarkably to the use of paints and lacquers. If a colored paint is applied to the back surface of a sheet of clear plastic, the paint adheres to the mirrow-flat surface of the plastic, and when viewed from the opposite side, the paint coat appears perfectly smooth and reflects light the color of the paint. If the plastic sheet is edge-lighted the paint will glow luminously because of the change of the angle of refraction of light through the clear plastic due to the paint clinging to the smooth opposite surface. This technique is being increasingly used in the manufacture of outdoor signs and similar applications.

SCULPTURING PLASTICS

Acrylic plastic blocks are excellent for sculpture work, Figures 48 and 49. They have no grain or natural cleavage planes, so the cutting work is greatly simplified, and since they are relatively soft, can be carved with ordinary hand tools. Plastic sculpture can be readily polished by ordinary polishing techniques. Especially beautiful effects may be had by using transparent, tinted acrylic carving blocks, and also with opaque and translucent materials in various colors.

The artistic possibilities of acrylic materials are almost unlimited.

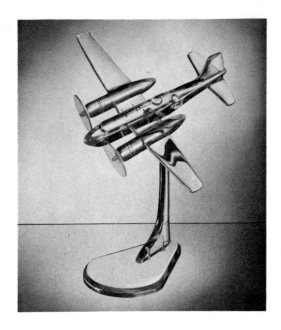

Fig. 49. Model carved from clear plastic by Floyd Dickey, South Bend, Ind.

POLYESTER RESINS AND REINFORCED LAMINATES

One of the most common commercial uses of plastic materials at present is in the manufacture of "laminates." This word comes from the Latin word lamina, meaning leaf. In plastics discussion, a laminate is a body built up of a succession of thin layers of some reinforcing material held together by means of one of the modern resins.

Since there are so many possible resinous or plastic materials, and so many possible reinforcing materials, the market today offers a very wide variety of laminated plastic-bodied articles. The plastic is one of the thermosetting resins, because with these, the material may be first applied in layers of liquid material with the reinforcing sheets built up, to form a body of any desired thickness. After the body has been built up, it may then be set by heating.

The properties of the laminate formed in this manner depend on the characteristics of both the reinforcing or laminating material and the plastic used. Plastics may usually be made as hard and stiff materials, or they may be plasticized to make soft and often stretchy materials with a consistency much like rubber. If the materials are hard and stiff, they are usually brittle and easy to break or shatter. If they are soft and stretchy, they are generally easily pulled apart, and have low tensile strength. Even the harder forms usually have comparatively low tensile strength.

Because of this fact many applications of modern plastics call for the addition to the plastic material of something that resists pulling apart, and to which the plastic itself may cling. We find a similar situation in the case of ordinary concrete. It has excellent ability to keep from being crushed (compressive strength), but very poor ability to resist a pulling force. In order to strengthen concrete in this direction we imbed steel rods, or wire mesh, which have much higher tensile strength. The reinforcing materials are so made that the cement clings tightly to them and whenever a pull comes the steel reinforcing takes the pull and the concrete holds together.

In the same way we can add materials with a higher tensile strength than the plastic, and the resulting reinforced plastic will be stronger. We usually make our plastic articles with thin sections, the most common type of reinforcing materials involve materials that are made of small fibers. For example, cotton cloth is made up of interwoven fibers of cotton. This fibrous material has considerable tensile strength which it imparts to reinforced plastic. There are many other fibrous materials that have been tried as the reinforcing element in built-up plastic articles.

FIBERGLAS

One type of fiber has proved to be very useful in many plastic applications. This is a fiber made by drawing out molten glass into long thread-like fibers. A glass fiber has a very high tensile strength for a given cross-section. These fibers are quite brittle when bent, but they will sustain a comparatively large load when that load is only a lengthwise pull on the fiber.

These glass fibers are spun into thread and woven into cloth to make a product sold under

Polyester Resins and Reinforced Laminates

the trade name of Fiberglas. This is the registered trademark of the material manufactured by the Owen-Corning Corporation.

Fiberglas may be obtained in the form of woven cloth, in the form of sheets, or rolls of matted glass fiber, and also as loose chopped fibers. The cloth is used where the greatest strength is needed. The matted fiber sheets and rolls are used to build up a thick layer, where strength is not so important, and the chopped fibers are used for filling in small areas, as will be described later.

Glass fiber materials have very high tensile strength. They also will stand up under a lot of flexing or bending, and will take hard blows well--that is they have "impact strength." They are not rigid--that is, they have little "structural strength." This latter weakness is taken care of when the fibrous glass material is imbedded in plastic, while at the same time these fibers lend to the plastic, in a greater or less degree, their extraordinary tensile, flexural, and impact strength. The plastic material, applied in a liquid state, completely surrounds the glass fibers individually, and clings tightly to them.

Fig. 50. After sanding the bare wood to a smooth finish, the Fiberglas cloth is stapled or tacked in place, and the first coat of polyester resin is applied.

Fig. 51. Fiberglas cloth is stretched into place over half the boat, then squeegee is used to smooth it out and work resin through cloth mesh.

As a result of this, whenever pulling or tension strain is put on the plastic, the glass fibers resist stretching or breaking, and add tremendously to the tensile strength of the laminated material. The same thing is true to a lesser extent for resistance to sharp blows and to bending strains.

There are two principal types of plastic resins that are being largely used at this

Fig. 52. After second half is covered with Fiberglas, application of another resin coat helps to give complete saturation and smooth finish to boat.

time for laminating fibrous glass materials. These are the Polyester resins and the Epoxy resins. The Epoxies are able to stick tightly to metals, whereas the Polyesters are not.

POLYESTER RESINS

Because of the fact the Polyester resins are at this time less expensive than the epoxies we shall speak mainly of the polyester resins, except in the few instances where it is necessary to bond the laminated product to metals.

Many of the thermosetting resins "set" under heating, with a reaction that gives off gases. For this reason, they have to be set in rigid molds under considerable pressure. If this is not done, the gases released during the reaction tend to form bubbles or other irregularities in the final product.

WET LAY-UP METHOD OF PLASTIC LAMINATION

Fortunately the polyesters (and the epoxies) are not subject to this type of reaction. They set without any gas being given off, and consequently may be handled without using expensive molds and high pressures. This enables us to carry out what is known as the "wet lay-up" method of plastic lamination.

The wet lay-up method of preparing glass fiber laminates with plastic is such a simple job that anyone can turn out expert work with very little practice and technical knowledge. Even the amateur, after a little preliminary practice, can make such complex things as "plastic" car bodies or glass fiber boats.

The first requirement is to have some sort of a shaped surface upon which we can "lay-up" the particular object we wish to build. There are two general types of molds-- one type is so made that the object is built up on the outer surface of a more or less convex object, the other is so made that the object is built up on the inside of a more or less concave surface. We shall refer to them as "outside" and "inside" molds.

For example, in the chair shown on page 253, it is easy to see that we want the smoothest and best appearing surface to be that on the inside, or concave surface of the chair. With this in mind, we can immediately tell that we want an outside type of mold--one that is properly convex, and so shaped that the glass fiber and plastic can be laid up over this protruding surface. In the case of the overnight case, page 251, we want the best appearing surface to be on the outside of the box. Therefore we must prepare a hollow mold

Fig. 53. Left. Trimming Fiberglas to rail line. Fig. 54. Right. Using belt sander to smooth lumps or seams. Added coats of resin give a hard glass-like highly glossed surface. Color pigments may be added to resin in any or all coats.

and build up the box on the inside of this mold. We can discuss the actual preparation of the mold under the heading of the particular project.

Meanwhile, we will discuss the method of wet lay-up in general terms and find out what must be done, in all cases, no matter what may be the size and shape of the mold, or the object being made. Refer to Figures 50 to 55 incl.

The first step is to prepare the surface on which the laminate is to be built up. If this job is to prepare a metal surface so plastic may be bonded to it, the first step is to remove all paint, dirt, scum, and grease and get down to the base metal. Using a disc sander with a coarse sanding disc is best. Leave the metal somewhat rough so that it has some "tooth" for the resin to bite into. A similar sander is also suitable for preparing wood or other surfaces. Clean the sanded, bright surface with a hydrocarbon solvent, such as VM&P naphtha, to remove all traces of oil or grease. Oil or grease on a surface will prevent the proper bonding of the plastic to it.

If you are building something from a form or mold, then the surface of the form must be so prepared that the plastic will NOT stick to it, and can be removed quickly and easily after the plastic laminate has set. In case of wood or plaster molds or forms, two steps are required. First, a mold <u>sealer</u> must be applied to fill the pores in the mold. This may be done by applying clear lacquer, by brushing or by spraying. The sealer should be applied in at

Fig. 55. The finished product.

least two coats and allowed to dry completely before further work. After the sealer is dry, add a <u>parting agent</u> or <u>release agent</u> -- something that will prevent the plastic from sticking to the surface of the form or mold. There are a number of these on the market under various trade names. Any good car or floor wax may be used. The parting agent should be applied evenly, in a thin coat, over the entire mold surface--then buffed to a high gloss to assure evenness of surface application.

The second step, after the lay-up surface has been prepared is to measure and cut all pieces and strips of Fiberglas cloth. This is important. The "pot life" of the resin mixture, after the catalyst and accelerator have been added, is only about fifteen minutes to a half hour. Unless the glass cloth is cut and ready to apply before mixing up the resin mixture, you have little time left for doing it afterward. Plan ahead. Cut all the glass cloth you will need to go with the amount of plastic you plan to mix in the first batch--then lay these pieces out on the floor or workbench within easy reach.

The next step is to mix the resin with the required catalyst and accelerator, following the manufacturer's specific instructions. Use the catalyst and accelerator supplied with the resin you buy, or that which is recommended by the manufacturer of the particular resin you are using.

Some catalysts work better than others with a given formulation. The catalyst and the amount of catalyst may vary according to the thickness of the layers of resin-glass material.

The catalyst is first mixed with the resin, mixing slowly and carefully to assure thorough mixing without the inclusion of air bubbles, which may be hard to remove. Next the accelerator is mixed in the same fashion, and the material is then ready to apply--and must be applied within the next fifteen minutes or so. <u>Never mix the catalyst and accelerator together first</u>--some such <u>mixtures are violently explosive</u>--and in all cases superior results may be had with the two-stage mixing.

As soon as the resin-catalyst-accelerator mixture is ready the actual lay-up work must be started. There are two ways to saturate the glass fiber cloth with resin. One obvious way is by placing the cloth pieces in the resin. For this operation, the prepared resin mixture is poured in a flat pan of suitable size, and the glass cloth pieces are dipped into this to saturate them with the resin before they are applied to their proper place in the lay-up. If you are using epoxy resins, wear rubber gloves during the dipping, since epoxy materials are irritating to some people's skin.

APPLYING RESIN WITH BRUSH

Another method of applying resin to the glass cloth is to paint it on with a brush. This method is recommended when working with larger surfaces, as when making a boat hull or auto body. In this method, paint a layer of the resin onto the surface of the mold or form--lay the glass cloth on this freshly painted area--then apply one or two coats of the resin to the top surface of the glass material with the brush. Brush the resin in carefully being sure to get complete saturation. Unsaturated areas will show up white, and must be soaked in order to get a good job. Smooth out wrinkles, and work out all air bubbles in the mixture as these may show up as pits in the finished work, and will certainly weaken the bond between the glass fibers and resin at the point of the bubble. Bubbles may be worked out with the hand or brush, or with a squeegee. In some instances it may be advisable to lay a sheet of cellophane or Saranwrap over the area where the bubbles are, and then rub over the surface of the outer film to remove the bubbles.

Multiple layers may be applied one on top of the other without waiting for the lower layers to set. To gain greater strength in the final lay-up, the usual practice is to lay the glass cloth on so the long weave of the cloth is at right angles to that of the piece just below it.

As soon as the glass has all been laid on, apply a final thin coat of the resin over the entire outer surface and smooth out carefully--then allow the lay-up to harden. Both the polyester and the epoxy resins will harden at ordinary temperatures, when mixed with the proper catalysts and accelerators. The hardening may be speeded up by applying heat--especially by directing the rays from heat lamps evenly over the surface. This latter procedure is not recommended for large jobs, such as boat hulls or auto bodies. The heat must be applied evenly to get an even cure. Hardening of the plastic lay-up is slower in cold, damp weather, so allow a longer time in such cases. In hot weather, if the job is exposed to sunlight or heat-ray lamps, the resin will harden in an hour or so. It is hard enough for final finishing when it cannot be gouged with the finger nail.

If your job calls for laying the glass cloth and plastic onto a vertical surface, there are certain materials called "thixatropics" which may be added to the resin and cause it to jell or become very thick so it will stay in place better. Another method of holding the materials in place is to cover with cellophane or Saran Wrap which can be tied or taped in position to hold the glass cloth and resin in place until it is hardened.

The final step is finishing the surface after it has been removed from the form or mold. For finishing, the surface must be hard enough that it cannot be marked with a fingernail. Smoothing the surface can be done in minutes using a disc sander with a No. 16 or a No. 24 grit sanding disc. A body file or a block sander may be used for final touch-up work.

If the foregoing procedure is followed, even the amateur can, with a little practice, produce excellent results. Perhaps before you try any actual project, and just to get the "feel" of working with Fiberglas and plastic resin, you might stick a patch of the glass cloth onto a board with a small amount of the plastic resin. Use a small square of convenient size, say 3 or 4 inches. Clean an area on the board somewhat larger than the size of the glass cloth square and wipe away all sawdust. Paint on a layer of the plastic material to which the catalyst and accelerator have been added, place the glass cloth square on this wet plastic area, and paint more of the plastic over the upper surface of the glass cloth, working it in until all white areas disappear and you can see through the glass and plastic at all points. Allow to harden. Finish by sanding and waxing. You will notice that if the glass fibers have all been saturated, the entire patch will be transparent and the grain of the wood can be seen through it as though the wood were covered with a clear varnish or lacquer.

The plastic material can be dyed almost any desired color, or can be made opaque by adding the proper coloring or coloring materials to the liquid plastic before adding the catalyst and accelerator.

APPLYING PLASTICS TO METALS

If your job calls for applying plastic and Fiberglas to metals, be sure to use epoxy resins. Polyester resins will stick to almost any clean, non-oily surface except a metal. Epoxy, on the other hand will stick to ANYTHING that is not oily or greasy.

One of the most spectacular uses for epoxy resins today is in patching sheet metal, such as automobile bodies, metal tanks, etc. In general, almost any hole, rent, dent, or tear in a piece of sheet metal may be repaired with glass fiber cloth and epoxy resins. The metal surface must be carefully cleaned of all paint and grease, and sanded to a clear surface, leaving it a little rough to provide "tooth" for the resin. In the case of large holes or rips, it is well to back up the opening with a layer or two of plastic and fiber on the wrong side, then lay saturated Fiberglas cloth into and over the hole, covering the edges of the hole by at least 3 or 4 inches onto the good metal. Apply epoxy resin with the glass cloth as outlined in the foregoing paragraphs, allow to dry, sand down, prime and paint. The patch made will be at least as strong as the original sheet metal itself.

The following projects will acquaint the experimenter with the basic principles of using Fiberglas-plastic lay-ups.

FIBERGLAS-RESIN SERVING BOWL

SUBMITTED BY:
 Charles Alm
 Lafayette High
 Lafayette, Indiana

This is a very unique project in Fiberglas-Polyester resin materials. It is easily made. Not only is it attractive and desirable, but a useful item. Sets of bowls may be made.

MATERIAL REQUIRED:
 (for one 10 in. bowl)
 1 sq. ft. light weight Volan finish Fiberglas cloth
 8 oz. Polyester resin with preadded Nuodex
 1/2 oz. Lupersol DDM
 2 oz. Syloid #244 thickening powder
 1 sq. ft. medium weight drapery material (loose weave preferred - your choice of color and pattern)
 1 pt. Acetone (for cleaning)
 1 2 oz. jar P.V.A. - mold release

PROCEDURE:
1. The first step is to secure a mold for making the bowl. Male and female molds may be turned on a wood lathe. Hard wood such as maple should be used. Allowance must be made for the thickness of the final bowl in the making of the matched molds. This should be approximately

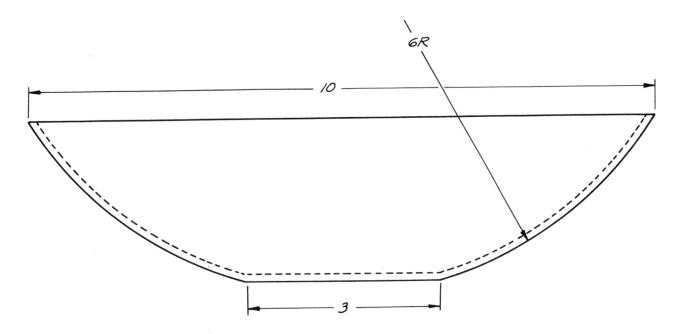

3/32 in. Sand surfaces very smooth.

2. A coat of clear Epoxy resin should be brushed on both halves of the mold to form a hard, nonporous, sealed surface.
3. Apply a coat of P.V.A. mold release to working surfaces of both halves of mold. Allow 15 minutes to dry.
4. Using an 8 oz. glass jar, mix about 4 oz. of the Polyester resin (half full) with the balance of the jar filled with Syloid #244 powder. Mix thoroughly, as this becomes the thickening agent that will prevent the resin from running into center of mold.
5. After mixture is made, add approximately 40 drops of Lupersol DDM to mixture. Mix thoroughly.
6. Brush mixture on female half of mold.
7. Immediately place layer of Fiberglas cloth into mold making sure it is completely saturated. Brush another coat of the resin mixture over cloth, fully saturating it.
8. Mix balance of resin in clean glass jar with approximately 30 drops of Lupersol DDM. Mix thoroughly.
9. Into the second layer of resin which was brushed on, lay the piece of drapery material, with design on top side.
10. Brush another layer of resin (new mixture) into bowl, making sure the drapery material is thoroughly saturated. It is suggested at this point that the resin be allowed to "puddle" in center of mold, assuring plenty of resin for next step.
11. Using male section of mold as a pressing fixture, press evenly into other section, making sure that resin squeezes out around all sides, thus assuring a completely filled mold. Hold in position by placing a weight on male half of mold.
12. Allow to cure for four hours, or more, to assure curing of part before removing.
13. When fully cured, remove from mold, and trim edges to even curve all around. This can be readily done on a band saw.
14. File or sand edge smooth, and polish on buffing wheel.

NOTE: Because of temperature and moisture conditions of the weather, a slight variation in the amount of Lupersol DDM may be necessary. This may be determined by experimenting.

It is best to select cloth (drapery stock) that is rather loosely woven, thus allowing plenty of "stretch" to conform to wood mold. Before beginning the layup operation, press the drapery stock into mold thus shaping it before actual operation with resin takes place. If resin is fairly thick in consistency, Syloid powder may be omitted. This will result in a clearer underside of bowl.

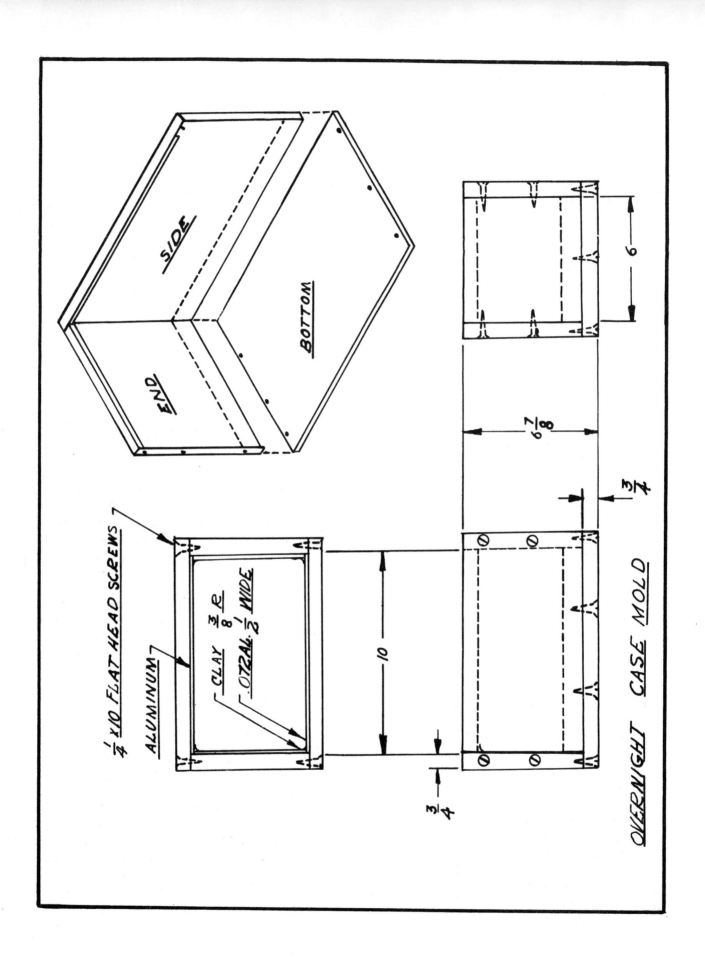

OVERNIGHT CASE
(FIBERGLAS-POLYESTER RESIN)

SUBMITTED BY:
Jack Berry
McAlester High School
McAlester, Oklahoma

This well made overnight case is a fine project in the Fiberglas-Polyester resin plastics. This requires a mold which will give the woodworking class a project. The mold, when completed, will be usable, over and over again, to produce these cases.

MATERIAL REQUIRED:

For mold: Approximately 4 sq. ft. 3/4 in. maple lumber
Small amount of plastic clay
24 - 1-1/4 in. #10 FH screws
Approx. 3 ft. of flat aluminum strip .072 in. thick x 1/2 in. wide

For one case: Approx. 1/2 lin. yard of single weight Fiberglas cloth, 44 in. width
1 quart Polyester resin (with preadded Nuodex)
1 bag, 4 oz. Syloid #244 thickening agent
1 oz. Lupersol DDM
1 paint brush (old or inexpensive brush)
1 pint Acetone (for cleaning)
1/4 lin. yard Fiberglas Mat

NOTE: Material quantities are specified certain amounts; however, one has to vary the amounts according to temperatures, weather conditions, etc. which have a slight reaction on the hardening time, and general working conditions. Amounts specified are ordinarily used at dry, moderate temperatures.

1. Two wood forms are prepared for molding the case. Sketches show the build up of the bottom mold. You will note that the mold has been put together with flat head screws, thus, allowing for disassembly of mold to release the finished case half. The mold or form for the top of case is made in same way as bottom mold except that mold is made only 1 in. deep and requires no aluminum stripping. The aluminum stripping which is fastened into the bottom half mold is fastened by use of flat head nails or small screws placed through pre-drilled holes. This aluminum stripping serves the purpose of forming an indentation at the top edge of the bottom section which allows lid to fit down over it.
2. Aluminum strips are screwed into place upon completion of the wood form.
3. Using a radius form tool, apply plastic

clay in corners and over joints, thus preparing the way for an even radiused corner and edge on the case.
4. Smooth inside surface of mold by sanding.
5. Apply coating of P.V.A. mold release to inside of mold, <u>thoroughly</u> coating all surfaces. It is very important that all surfaces be coated otherwise material will stick to mold and create marred or pitted surface.
6. Before mixing of resins, it is necessary to prepare cut Fiberglas cloth, to size, thus eliminating delays between coatings of the resin. Fiberglas cloth may be in several sections, or odd pieces left over from previous runs. Corners can be wrapped around easily, thus, this size cutting is left up to the student. The Fiberglas mat or woven mat should be cut to size also for reduced time loss while applying resins.
7. Resin preparation:
Mix only as much resin as can be used up in a short time before hardening takes place. DO NOT MIX THE ENTIRE AMOUNT OF RESIN AT ONE TIME. The first coat into the mold is called the Gel Coat since it is the one which we must be certain sticks to the side of the mold. Using about 1/2 pint of resin, mix an equal volume of Syloid #244 filler powder. This is a thickening agent which helps in keeping the resin from being too "runny." <u>Mix thoroughly</u>. Next, add approximately 30 to 40 drops of Lupersol DDM, <u>mixing thoroughly</u>.
8. Apply this gel coat to inner walls and bottom of mold. Allow to become tacky before proceeding.
9. Place one layer of Fiberglas cloth in position, covering all inside surfaces. Ends of cloth sticking out of mold is satisfactory, since it is trimmed off later to even edge.
10. Mix about a pint of resin with 40 to 50 drops of Lupersol DDM, mixing thoroughly. NOTE: This amount should be varied if needed (the number of drops) according to weather conditions. During damp weather, more drops may be needed. Test for best results. This is the mixture which will be used for the remaining coats of resin.
11. Apply coat of resin into mold over first layer of Fiberglas cloth. Be sure that cloth is fully saturated, and that no dry spots remain.
12. Place coat of Fiberglas mat (or woven mat) into place in mold.
13. Coat with another coat of same resin mixed in step #10.
14. Apply final layer of Fiberglas cloth.
15. Apply final coat of resin, being sure all is well saturated.
16. Allow to <u>harden in mold</u> overnight.
17. Remove by taking mold apart.
18. Top half of case is made in exactly same manner as bottom half however no aluminum strips are involved, and also, only 1 in. depth is required.
19. If a permanent color is desired in the makeup of the case, color paste of a variety of colors may be added prior to painting on the last coat of resin. Check with your supplier as to proper amounts needed, according to color.
20. Upon removal from mold, small voids or pits may be found in the surface of the case. This can be the result of several causes. One might be the fact that the mold release was not covering the surface fully; possibly too, they are small gas pits caused by over-catalyzation in which case reduce the number of drops of Lupersol DDM in step #10.
21. After completion of case top and bottom, an additional coat of resin may be applied (mixed similar to step #10) to help fill up the small pits or voids. Apply with brush and allow to harden. Before applying resin, use fine sandpaper to remove film of grease (mold-release) from case halves, which will also provide a good surface for final coat of resin to adhere to.
22. Case top and bottom are matched and hinges, latches, handles etc. are put on with regular metal fastenings. Case may be drilled easily with regular carbon or high speed drill bit.
23. For final finish, wax with any good paste wax.

FIBERGLAS-POLYESTER RESIN CHAIR
(Modern Bucket Type)

SUBMITTED BY:
 Clyde Brown
 Carthage High School
 Carthage, Illinois

Constructed of Fiberglas and Polyester resin, this chair is attractive, and comfortable. The legs which are constructed of wrought iron can be bolted to the seat, or they may be constructed as part of the seat itself by use of Fiberglas strips laminated on with Polyester resin. A variety of colors can be made by the addition of color pastes in one of the final applications of resin. A mold which is required can be made by using a wood frame covered with chicken wire, and then covered with plaster and smoothed to shape. Or, a plaster cast can be made from a commercially purchased chair.

MATERIAL REQUIRED:
(Sufficient resin and cloth are listed to adequately make one chair of average size with allowance for waste)
- 2-1/2 yards x 44 in. wide of light weight Fiberglas cloth, Volan finish
- 1 yard x 44 in. width, Fiberglas mat (since this is generally sold on approximately a 72 in. width, it may become necessary to purchase a yard of this by the width available).
- 1 gallon Polyester resin with preadded Nuodex
- 1 pint Lupersol DDM
- 2 oz. Color Paste (of desired color)
- 1 quart Acetone (for cleaning)
- 1 pint P.V.A. Mold release solution
- 2 wrought iron legs 5/8 in. diameter x 4 ft. long
- 4 No. 10 angular leg Domes of Silence
- 1 dispensable paint brush (about 2-1/2 in.)
- 4 oz. Syloid #244 powder

PROCEDURE:
1. Obtain or make form as indicated. Male form is preferable.
2. Paint on a layer of the P.V.A. release solution, and allow to dry.
3. Give form a coat of wax to insure release (such as Johnson's Paste Wax).
4. Lay out all materials for making chair, cutting such items as the Fiberglas cloth and mat to slightly larger sections than needed. In all cases, piece units of materials can be used and lapped together as this will result in a saving of cloth and mat, and can only be detected in rare instances. This, too, can be surfaced with

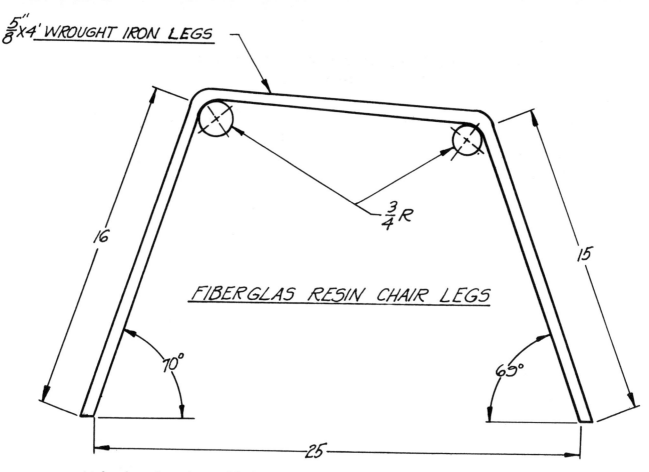

a coat of colored resin, and hidden from view.
5. Form wrought iron legs as indicated and assemble Domes of Silence.
6. Mix about 8 oz. of resin with 8 oz. volume of Syloid #244 powder. This is a thickening agent which helps to prevent "run" of the resin to bottom of the seat, or over outer edge. Add Lupersol DDM to mixture (about 100 drops).
7. Brush mixture over form.
8. Place first layer of Fiberglas cloth in resin over form.
9. Mix another 8 oz. resin with Lupersol DDM (leaving out the Syloid).
10. Brush the mixture over cloth, thoroughly saturating it. It is thoroughly saturated when cloth appears clear throughout.
11. Place layer of Fiberglas mat in position, and thoroughly smooth out. Sharp contours that cannot be formed, can be covered by using small pieces cut and placed to fit.
12. Mix 1-1/2 pints of resin with proper amount of Lupersol DDM.
13. Brush mixture over mat, again thoroughly saturating it.
14. Place second piece of Fiberglas cloth in mixture, and smooth in place.
15. Brush on resin mixture (from step 11 above) giving thorough saturation.
16. Allow to stand until dry to touch.
17. Pull resulting chair seat from mold.
18. Mix 1 pint of resin with desired color of paste until proper shade of color is obtained. Mix in required Lupersol DDM.
19. Brush mixture onto one side of chair, and when dry, brush onto opposite side.
20. Trim edge of chair and sand smooth if necessary.
21. When dry to touch, place leg in position and place strips of Fiberglas cloth over leg as a "hold down" in several places, or along entire contact area, and brush on a coat of uncolored resin mixture Follow with additional coat of colored resin mixture and allow to dry.
22. After dry, sand with #400 wet-or-dry paper to smooth surface.
23. Give entire chair a coat of uncolored resin mixture thus giving a clear finish above all colored surfaces. Allow to thoroughly dry.
24. Apply a coat of good furniture wax.

OTHER APPLICATIONS OF PLASTICS

So far we have discussed plastics largely from the standpoint of their chemical makeup and their handling and working in the home and school workshop. In this chapter we will discuss the more common commercial methods of fabricating plastic articles and some of their applications.

When making plastic materials for the market, many angles must be considered. It may be cheaper in the long run to use a more expensive plastic raw material that can be manufactured by automatic machines than it would be to use a cheaper raw material that requires expensive handwork, or long periods of using valuable equipment.

Since most of the plastics are, at one time or another in a truly plastic condition, this plastic or moldable material can be put into a mold and molded into various shapes. Therefore molding is one of the most common techniques found in commercial use of plastics.

TYPES OF PLASTIC MOLDINGS

Commercial plastic moldings are of two chief types--compression moldings and injection moldings. Basic plastic materials as furnished by the plastic manufacturer to the fabricator, come in the form of molding powders or granules, and also in the form of liquids, sheets, films, rods and tubes.

MOLDING PROCEDURE

In the case of thermosetting plastics, the molding powders or granules are usually in the second stage of development--that is, they are ready for the final heating and curing or setting under heat. Thermosetting materials are normally in their final chemical state, but are plasticized by heat, forced into the molds, and allowed to cool in the molded form.

COMPRESSION MOLDING

Thermosetting plastics are largely molded by the compression molding technique. This is because these materials are usually quite viscous during the final heating stage, and also because most thermosetting materials give off hot gases or vapors during the final cure and need to be under pressure during the "setting" stage so the bubbles of escaping gases will not distort them or make them change shape.

In its simplest form, compression molding is done with a two-part mold that is divided along some natural division line of the completed product. The mold is first heated (usually to a temperature of about 300 deg. F.) and then filled with a carefully measured amount of the molding material. This molding material consists of plastic molding powder or granules together with a filler--perhaps wood flour, high-grade cellulose, mica or other mineral matter. Fillers are of a wide variety and are added to strengthen and to fill out the plastic, and also to give certain desirable properties to the finished product.

In the typical molding operation the mold is heated, filled with the mixture of plastic granules or powder plus filler, all weighed or measured out to exact proportions. Then the mold is placed in a dydraulic or other suitable press and subjected to a pressure of about 2,000 pounds per square inch. Heat is then applied to the mold and its contents until the resinous content is set or cured. This latter time may take from a few seconds to an hour or more. There are many variables in the molding process.

During the curing period in the mold, thermosetting resins undergo a chemical change and are hardened. Usually they can be removed from the mold while still hot, thus saving the time and expense of having to reheat the mold for the next operation.

One variation of compression molding is called transfer molding. In this type of work the thermosetting plastic material is first heated in a transfer chamber. This heated material is then forced into the mold through a passageway from the transfer chamber, and is allowed to set and harden in this mold. This sort of molding is often used where metal inserts, such as in electrical devices, are required. These are put in place in the mold chamber and the fluid plastic from the transfer chamber flows around these parts, then hardens with them in their proper place.

INJECTION MOLDING

The second over-all class of commercial moldings, injection molding, Figures 56 and 57, was developed largely to speed up the job of making moldings of thermoplastic materials. In this method of molding, the plastic is heated in a separate heating chamber, and then, while hot and semifluid or fluid, is forced into the mold by means of a hydraulic ram or similar device. Injection moldings can be made in multiples--the mold may hold numerous cavities that are made of the shape of the desired finished piece. These cavities are all connected to the chamber containing the molten plastic. When the press plunger forces out the melted plastic, it fills all the cavities at one time. Since the mold is not heated, the thermoplastic materials begins to set immediately upon contact with the cold walls of the mold. In a very short period, the plastic is set (by cooling) and the mold may be opened and the molded pieces removed.

The injection molding process is suitable only for rather small articles that are rela-

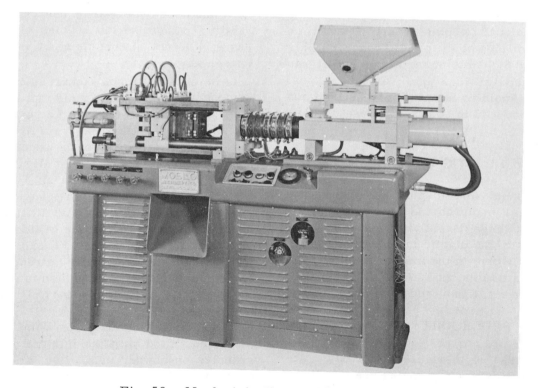

Fig. 56. Moslo injection molding machine.

tively light in weight. See Figure 58. In general, large moldings, even of thermoplastic material are better made by compression molding.

In practically every molding operation there is some protruding material at some point on the molded piece which marks the entrance point of the fluid plastic (the "gate") or the parting line of the two parts of the mold. This protruding material is called "flash" and must be removed, usually by hand, after the molded piece has cooled and before it is ready for further operations.

Molding is a complex operation and the molder must contend with many variables. Some materials, such as nylon, are very difficult to mold. Sometimes parts can be machined from blanks of the material more cheaply than they can be molded.

EXTRUSION MOLDING

Another method of forming plastic parts that is often classed with molding is extrusion. It is largely used in preparing rods, tubes, sheets, etc. from thermoplastic material.

Basically, the process of extrusion involves forcing the plastic material, while in a plastic or semi-melted condition, through an orifice of the desired shape. In most extrusion operations, the plastic material is fed into a hopper, which in turn feeds the material into a heated chamber where the material becomes plastic or semiliquid. From this chamber, the material is forced out by means of a mechanical screw through an opening that has been made to the shape desired for the extruded article. If a round rod is to be the product, the material is forced through a round hole. A tube can be extruded by having a metal rod extending out from the interior of the machine into the center of the round orifice--the extruded plastic flows out of the round slit between this rod and the outer circumference of the hole. A sheet may be made by extruding the plastic through a slit.

Certain formulations of some of the thermosetting plastics may be extruded while they are still in a "pasty" condition--the material is heat cured after extrusion.

Fig. 57. Stokes injection molding machine in operation.

Fig. 58. Plymouth parking light lens molded from Plexiglas.

BUILT UP LAMINATES

Another method of using plastics is in the building up of laminates, Fig. 59. Many other applications of the laminating principle are possible. For example, sheets of paper, cloth, or any similar material may be saturated with liquid resin, stacked in suitable thicknesses, and heat cured between the flat plates of a heated press to build up laminated sheets of any desired thickness.

Most of the materials on the market of the type mentioned here are made with phenolic resins, because of their relative low cost and excellent properties. Laminated sheets find many applications in manufacturing a wide variety of objects in everyday use. Tubular laminates, made by winding soaked sheets of paper, cloth, or other material on a mandrel followed by heat curing, also find many uses.

Another type of laminated material that we are all familiar with is plywood, which is much used in the construction industry and by the homeshop hobbyist. Most plywoods are made by building up a laminate of thin sheets of wood coated with liquid thermosetting resin. In many cases the thin wood sheets are made by peeling off a continuous "shaving" by rotating a suitably prepared log against a sharp knife. After these sheets have been flattened and dried they are stacked in layers with the grain in each successive layer lying at right angles to the one below it. The faces of the wood sheets are saturated with liquid resin as each layer is applied until the desired thickness is reached. The plywood sheet is then put between the plates of a heated press and the resinous bonding layers are cured.

In speaking of plywood, we might well mention again at this point another type of building board that uses the natural thermosetting material lignin as a binder. All woody matter is made up of cellulose, various members of the family of chemical sugars, etc., and all are bound together by a plastic-like material called lignin. If wood chips made from waste woody matter are strongly heated under pressure, with water, and the pressure is suddenly released, the material "explodes" under the pressure of internal steam. The cellulose fibers separate, become "fluffy," and the lignin separates from the fibers. The mass is treated to remove the sugar-like materials, and the rest is a moldable, thermoplastic material. The thick fibrous mass cannot be handled by injection molding. In compression molding, the materials will react with amines, furfural and phenol to give a thermosetting material. This is usually molded flat, between hot steel plates in a

Fig. 59. Hydraulic laminating press (Pasadena Hydraulics, Inc.) used for compression molding as well as heat laminating of thin plastics in "sandwiches." Pilot models of items can be compression molded in this machine which utilizes heat and pressure to mold the parts. The press develops up to 40,000 pounds platen pressure. Lamination or "heat sealing" of cards, photos, etc. can be done with this type press.

press to form the building board that goes by various names such as hard-board, pressed-wood, etc.

COLOR LAMINATES

Color laminates or "sandwiches," are usually made from phenolic sheets of different colors, laminated together. For example, we might have a sheet of opaque white material sandwiched between two thin sheets of black. If a piece of this material is then carved in such a manner as to penetrate the black outer layer, the white of the inner layer shows through. Laminates such as these are usually called engraving laminates, and are used for name plates, signs, and certain decorative purposes.

PLASTIC CASTING

The word casting has for a long time been used to describe the act of pouring molten metal into a mold, then allowing the metal to cool and harden and thus take the shape of the mold. In plastic technology the word casting is used in somewhat the same meaning. In general the term indicates the formation of solid plastic materials in a mold formed by pouring a liquid into the mold and allowing it to harden, and thus assume a predetermined shape, without the use of pressure. In this respect casting differs from compression and injection molding.

In most casting of plastics, the process includes the polymerization of the resin within the mold--or at least the partial polymerization, as in the case of some of the thermosetting materials.

One application of the casting process that students of this book will find most interesting is the casting of acrylic sheet, rod, and tube stock--materials such as are used in the projects section of this book. Since Plexiglas is a thermoplastic material it is possible to extrude it. The heat-softened acrylic is, at its softest, a highly viscous material that will flow only under considerable pressure. This slow moving flow tends to set up strain lines in the plastic, and these strain lines disturb the optical qualities of the finished material.

Since one of the greatest appeals of Plexiglas and other acrylic plastics is their beautiful optical clarity, every effort is made in manufacturing them to preserve these optical properties. One method of doing this is to cast the plastic in molds made of highly polished, smooth plate glass. Briefly this process (if a sheet of acrylic is to be made) calls for polymerizing the acrylic in place-- that is, in the mold. Two sheets of fine plate glass are separated the proper distance to make a finished sheet of the desired thickness, and a suitable gasket is inserted around the edge of the opening to contain the liquid material to be poured into this mold. It is possible to use a solid gasket, such as aluminum, or to use a rubber-like material such as plasticized polyvinyl chloride that has been extruded in a ribbon of proper thickness.

During polymerization of the monomer, there is a considerable shrinkage. If solid gaskets are used around the edges of the glass sheets that form the mold, they must be removed soon after polymerization begins, to allow the glass sheets to move inward with the shrinking acrylic. If the "squeezable" PVC ribbon is used, it will compress with the shrinking acrylic and allow the glass sheets to remain in contact with the two faces of the acrylic sheet being formed. Since the finished plastic sheet faithfully reproduces the level, polished surface of the glass sheets forming the faces of the mold, we can easily see what would happen to the surfaces of the acrylic sheet if it were allowed to pull away, irregularly, from the glass mold.

Another factor in casting acrylic resin is the sharp exothermic reaction of the polymerization--the reaction gives off heat at a certain stage. If not properly controlled this heat given off can shatter the glass molds and ruin the batch. For this reason, the casting molds must be placed in an oven that has positive air circulation. This heating factor, its relative uncertainty, and the difficulty of controlling it all contribute to the cost of making acrylic castings.

The mold shrinkage, which may run as high as 15 to 20 percent, takes place evenly in all directions. After the plastic has become very thick and viscous so that it cannot move easily in the mold, most of the shrinkage is confined to the smallest dimension--that is, to the thickness of the sheet. This uneven shrinking sets up stresses in the plastic which would interfere with its optical properties and its workability. Heating the sheet in an annealing oven to about 300 degrees F., and then allowing it to cool slowly and gradually removes these stresses almost completely.

Rods and tubes of acrylic are cast in a manner similar to the method just discussed for sheets. In each case the surface of the cast piece reproduces the smooth, even surface of the glass mold so that the finished product has the desirable smooth, shiny, polished surface that makes acrylic resins so attractive in appearance.

CAST PHENOLICS

Cast phenolics, which make up a good part of the commercial applications of cast resins, are of essentially the same composition as the molding resins, except that no filler is added to the cast pieces.

The ingredients of the resin--phenol, formaldehyde, plasticizers, dyes, and lubricants are first mixed and heated in steam jacketed kettles with very accurate temperature control. After about 24 hours, a viscous, resinous material is formed that has a syrupy consistency. This syrup is then poured into lead molds that are made by casting them about a polished steel piece which has been machined to the shape of the desired finished product. The lead molds are, as a rule, used only one time. Standard castings include sheet, rod and tube forms, and special castings are made for such applications as knobs and handles of various types, clock cases and so forth. Phenolic sheets are made in thicknesses ranging from 3/32 in. on up.

Cast phenolic resin in its finished form is nonflammable, tasteless, odorless and has excellent insulating qualities. The material may be machined readily, and worked with ordinary woodworking tools. Worked pieces when sanded and buffed achieve a high polish, and have an appearance of depth of surface that is attractive. The dimensional stability of phenolics is very good, and they resist

Fig. 60. Pouring Castomold over original gear to produce a mold.

many chemicals and solvents.

Phenolic castings that have been properly catalyzed may be polymerized at room temperature, or by controlled, low heats in an oven.

OTHER RESIN CASTING

Polyester and epoxy resins have been used to some extent in casting work. The casting of epoxies together with the phenolics castings are being used to a surprising extent in making draw dies for metal forming, together with other metal forming tools. Epoxies and polyesters are also being used to an increasing extent for casting electrical components.

CASTOLITE CASTING

At this point, it might be well to discuss a specialized type of polyester casting that has been introduced for the school shop, and home hobbyist, by the Castolite people.

Other Applications of Plastics

The Castolite Company has introduced a special formulation of polyester resin, together with the required catalyst and accelerator. Castolite is a formulation of polyester resin, together with catalyst and accelerator, that has been prepared for small lot use.

This material has been especially prepared for home or small shop casting use. It enables experimenters to prepare thousands of different types of castings, using a clear resinous material. Various types of specimens may be mounted in blocks of clear Castolite to preserve their original appearance and properties, and to permit continued viewing of the materials, without damage to them. Figures 60, 61, 62, 63. Mineralogical, biological and many other kinds of interesting and exciting materials may be imbedded in this clear plastic to give a three dimensional view of the specimens so mounted.

Castolite may be obtained from various sources, together with the literature describing step-by-step procedure for using it.

PLASTIC FOAMS

We have already mentioned the rigid plastic foam called "Styrofoam," and mentioned some of the possible uses of this material. Other plastic foams are also appearing on the market, many of them in the form of elastomers similar to foam-rubber. The elastomeric forms are being used wherever a cushioning effect is needed. Shaped forms urethane plastics are being used in the newer cars for making padded dashboards, etc. Rigid forms of foamed plastics are being foamed in place for use as insulating material to prevent the passage of heat.

Both thermoplastic and thermosetting resins are capable of forming foams. The method of producing the bubbles of gas that make the material vary greatly. Basically, the process is similar to that used in making biscuits rise in the oven. The foaming process must be carefully controlled to obtain the correct amount of bubbles per given quantity of plastic.

Rigid foams are finding much use in structural materials. The foamy plastic is foamed

Fig. 61. Removing hardened Castolite gear from flexible mold.

Fig. 62. Botanical specimens embedded in Castolite resin, are preserved for future study.

in place between thin strong skins of metal or polyester glass sheet. Panels made in this manner are suitable for many structural purposes as well as providing heat insulation.

A different, but somewhat similar use of plastics is in the construction of forms consisting of an inner core of honeycombed material--paper, different kinds of plastic films, etc.--faced on the outside with a rigid sheet of polyester-fiber glass construction. Radomes, usually hemispherical in shape, and used to protect radar equipment on shipboard or in radar equipped planes, are usually made of this material. Material of this sort is readily permeated by the ultra-short electromagnetic vibrations of radar equipment. This business of being permeable to electromagnetic waves must not be confused with the ability to conduct an electrical current.

Fig. 63. Novelities embedded in Castolite.

SHEETS AND FILMS

Almost all of the plastic materials can be produced in the form of thin sheets or films. These items vary from a few ten-thousandths of an inch in thickness, up to rigid sheets with thickness measurable in inches.

Many of the very thin films are used as wrapping materials to keep moisture in or out. Most of them are transparent and allow visual inspection of the wrapped goods without danger of contaminating the wrapped material by touching it or by entry of outside matter. Cellophane, a modified cellulosic material, was one of the first such films produced. Polyvinylidine chloride films, are sold under the trade name of Saran Wrap. Many plastic films are appearing on the market and have been found excellent for one or another special characteristic.

Still another, and rather recent development has been the bonding of plastic films to metal foil, thus giving a wrapping material with the excellent properties of both types of material. Polyvinyl film that has had a suitable amount of carbon black incorporated into it is electrically conducting, and may be used for electrical heating providing the film temperature does not go above 110 degrees F. The outlook for applications of plastic films appears almost unlimited.

Thicker films or thin sheets of transparent or translucent plastic have found use as lamp-shade materials. Some of these are made by bonding sheets together with decorative items placed between the layers of film to provide the decorations.

One of the rather spectacular uses of film is found in the strawberry industry. Polyethylene sheeting is spread along the row of strawberries. The berry plants are planted in the soil beneath through openings in the film. Water reaches the plant roots through the soil from the side, but the berry plant itself grows up through and rests on the polyethylene sheet. Berries so produced do not come in contact with the soil and are clean and free from earth rot, mildew and other afflictions so often found in such fruit. The polyethylene film spread over the ground in the immediate vicinity of the plant prevents the growth of weeds and grass, and keeps the berry row clean. This is only one example of the possible use of films from an agricultural standpoint. Plastic films and sheet have also found considerable use in glazing greenhouses and in other applications where it is desired to

Other Applications of Plastics

admit light but to hold in heat. The plastic materials are not so brittle and easily broken as glass enclosures, and hail damage which is so often disastrous to a glazed enclosure is kept to a very low minimum when plastic films and sheets are used. Many plastic formulations will admit more of the useful light of the sun than will ordinary glass, particularly in the ultraviolet range.

FILAMENTS

Filaments of plastic materials are finding increasing use in the manufacture of cloth and clothing. In many instances these threads have entered into active competition with the older thread making materials found in nature such as cotton, wool and silk. One of the first plastic filaments to become well known was DuPont's nylon which was introduced as a material for weaving women's stockings, and as a direct competitor for silk. Another was rayon which is a modified cellulose material--usually the "acetate"--although other modifications are also much used.

Others that have achieved wide popular acceptance in the textile field go under the trade names of "Orlon" (polyacrylonitrile), "Dynel" (co-polymer of acrylonitrile and vinyl chloride), "Acrylan" (polyacrylonitrile modified with vinyl acetate), "Saran" (polyvinylidine chloride and vinyl chloride), "Dacron" and "Terylene" (polyester fiber), and a great many others.

These synthetic fibers, together with other synthetics made from natural proteins such as the zein from corn, and protein from casein which is obtained from milk, are all being used in the manufacture of textiles that are on the market today. The field is unlimited.

METALLIZING PLASTICS

Another and comparatively new field for plastics is that of metallizing. With the proper apparatus, plastic articles can be covered with a metallic film, thus giving them the appearance of metal while still retaining the lightness and other characteristics of the original plastic. Materials may be coated with a metallic layer by one of four processes; electroplating, chemical reduction of metallic compounds in contact with the surface to be metallized, molten metal spraying, and vacuum metallizing. The last named process lends itself most readily to the coating of plastics.

As may be seen from the foregoing brief discussion of the applications of plastics, the uses of these materials are growing rapidly. In this book we have touched on only a relatively few of many possible uses of plastics in their many forms. One thing that should be evident to the reader is that plastics are still in their upward surge of growth. Many employment and business opportunities are available for men and women with a thorough knowledge of plastics, their uses, their fabrication, and their limitations.

GLOSSARY OF TERMS

ACCELERATOR: Chemical material that will speed up a chemical reaction. Example benzoyl peroxide, added to the polyester resins.

ACID: A rather large classification of chemical substances. Acids, when dissolved in water, (1) taste sour, (2) turn blue litmus red, (3) react with active metals, giving off hydrogen, and (4) neutralize bases. They do all this because they contain an ionizable hydrogen atom which carries a positive electrical charge. Organic acids, such as the acetic acid in vinegar, or acrylic acid are true acids and have a sour taste.

ACRYLIC: (Acrylic resins) These two terms are usually used to designate polymerized methyl methacrylate resins such as Plexiglas and Lucite.

ALKALI: A basic substance, that gives an alkaline reaction with litmus, turning it from red to blue. Alkalis are usually water solutions of hydroxides of sodium or potassium.

ALKYD: Word obtained by taking the first syllable of alcohol and the last syllable of acid, and changing the c to k and the i to y. It refers to resins polymerized from a mixture of alcohols and organic acids.

ANHYDRIDE: Literally, without water. A molecule that has lost two hydrogens and one oxygen (the elements in a molecule of water). In the sense used in plastics chemistry, it usually means a molecule containing the group $-\underset{O}{C}-O-\underset{O}{C}-$.

AROMATIC: Compounds that contain the benzene ring as the major part of their molecular structure.

ASPHALT: Black, gummy residue left after distilling crude oil by boiling off the lighter weight fractions such as gasoline, kerosene, lubricating oils, etc.

ATOM: Smallest possible particle of one of the elements found in the crust of the earth.

BAKELITE: First member of our modern family of synthetic resins to be developed. It is formed by the co-polymerization of phenol and formaldehyde.

BENZENE RING: A molecular structure made up of six carbon atoms joined to each other in a cyclic or ring formation, and with one atom of hydrogen attached to each of the carbons. It is usually represented by a simple six sided figure when used in chemical formulations. Usually acts as an unbroken unit in chemical reactions.

CARBON: One of the most abundant of the chemical elements. It is represented in chemical formulas by the symbol C. Carbon is one of the principal components of all organic chemical substances, including most of our synthetic resins or plastics.

CAST - CASTING: In the sense used here, casting means to pour a liquid material into a mold and then to cause it to harden, taking on the shape of the mold.

CATALYST: Any substance which will affect the rate of speed of a chemical reaction without, itself, entering into the products of the reaction.

CELLULOSE - CELLULOSICS: Chemical name given to the fibrous material found in the woody matter of plants. Cotton fibers are one of the purest forms of cellulose.

CHARACTERISTICS: Features or traits that help to identify a material.

CHEMISTRY: Science of the study of matter, and of the atoms and molecules that make

Glossary of Terms

up matter. Includes study of the ways matter reacts under different conditions and the basic make-up of all types of matter.

COMPRESSION MOLDING: Process of forming plastic materials by forcing them into a mold under pressure and allowing them to set in the molded shape before removing the pressure.

COPOLYMER: If two (or more) different kinds of molecules join to form a large molecule, the new substance is said to be a copolymer. If only one kind of molecule is involved in this reaction, the resulting material is known as a polymer.

COTTON LINTERS: Short fibers of cotton that cling to the seeds when the cotton is ginned. These short fibers are later removed and used as a source of cellulose in making cellulose - based plastic - like materials such as Tenite CAB.

CURE: In this book the word cure means the chemical reaction that takes place when thermosetting materials undergo their final hardening under the influence of heat and pressure. It is used more or less interchangeably with set.

DIELECTRIC: Refers to any material that will resist the passage of an electric current. In other words, anything that is a good insulator for an electrical current. Almost all of the synthetic plastics are good insulators; they have high dielectric strength.

DIMENSIONAL STABILITY: Ability of any material to resist a change in dimensions or measurement under various conditions. For example a piece of wood is dimensionally stable under almost all conditions; a wad of chewing gum is not.

DIFFUSED: Light in which a straight beam of light has been broken up so that it travels in all directions. Light passing through a sand-blasted plastic surface is diffused. Light reflected from a surface with many small irregularities is diffused.

ELASTOMER: Rubber-like synthetic materials. Chemical name for substances that are usually called Synthetic rubbers.

ENCAPSULING: Putting in a capsule. In electrical and other work this usually means coating an object - such as a transformer, or even an entire electrical circuit - with a water tight, gas tight coating of non-conductive material.

ENDOTHERMIC: Endo means inside of, thermic refers to heat. Any endothermic reaction is one that "soaks up" heat during the reaction so that heat must be continually supplied from an outside source. Baking a potato is an endothermic reaction.

EPOXY-RESIN: A relatively new class of resins. They are similar in handling and in properties to the polyester resins. They are much more adhesive to metals and some other materials than the polyesters.

EXOTHERMIC: Ex means out of. Thermic refers to heat. An exothermic reaction is one that gives off heat. The burning of a match, or of fuel oil or gas is an exothermic reaction because heat is given off by the reaction.

EXTRUSION: Sheet, tube, rod, or other shape of material that has been formed by forcing a molten or plastic material through an opening of suitable size and shape, followed by solidification of the extruded piece either by cooling, cure or other chemical setting of the material.

FIBERGLAS: A trade name owned by the Owen-Corning Corp., for materials made of spun, woven or matted glass fibers that are formed by drawing melted glass out into fine filaments. Fiberglas cloth or matte is much used in making laminated structures, using polyester or epoxy resins as binder.

FILAMENT: A long and very thin, thread-like piece of material, such as a single cotton fiber or a single section of spider web. Most threads are made by twisting together filaments or fibers. Some mono filament threads are made from synthetic resinous materials such as nylon.

FLASH: A thin film of plastic formed on a plastic molding at the point where the two edges of the mold join.

GATE: The word "gate" as used in speaking of plastics refers to the opening in a mold through which the liquid or semiliquid plastic is admitted or forced into the mold.

GLOSSARY: A list of words, giving the definitions of those words. A glossary is usually placed at the end of a book and explains words used in the book that are special words or that have special meanings as used in the book.

IMPACT STRENGTH: Measure of the ability of a piece of material to withstand a sharp blow such as being struck by a hammer, or dropped on a hard floor.

INHIBITOR: Any substance which, by being present, will slow down or prevent a chemical reaction.

INJECTION MOLDING: Injection molding of thermoplastic materials consists of forcing the viscous, hot material out of a heated cylinder, by means of a plunger, into a chilled mold where it sets upon cooling to the form of the mold.

INORGANIC: Composed of matter other than animal or vegetable. Inorganic, as applied to chemistry, means that the chemical substance does not contain carbon - is not an organic chemical substance.

INSULATE: To prevent, or at least to offer great resistance to the passage of an electrical current, or of heat. An electrically insulating material may also be called a dielectric material.

LAMINATE: The word laminate comes from the Latin word lamina meaning leaf. In this book we use laminate to mean a product built up by bonding together several layers of thin or sheet-like material to form a much thicker layer or sheet.

LIGHT PIPING: Ability of some highly transparent materials such as optical glass and Plexiglas to pass light from one end to another of a polished rod with little loss of light, and even around bends or rounded corners. This is possible because the light is reflected back into the rod at the polished sides.

MACHINING: Process of shaping a block of metal or other solid substance by removing material from the block in the form of small chips or thin layers by means of a mechanical tool operated by a suitable power source.

MELAMINE-RESINS: Polymerized resinous materials formed by the reaction between formaldehyde and melamine, a nitrogen containing organic molecule much like urea, but more complex in its make-up.

MOLD: A mold (British spelling, mould) is a hollowed-out form that transfers its inside shape to any liquid-like material that is forced into it and allowed to set.

MOLECULE: Smallest possible particle of any given material. Most molecules are made up of more than one atom - usually different kinds of atoms. Some molecules, such as the polymerized resins, are almost unbelievably complex.

MONOMER: Single molecular unit that enters into a reaction with other similar units to form a polymer. Mono (one) mer (part) - poly (many) mer (parts).

OPTICAL PROPERTIES: Properties of a substance that deal with its ability to transmit or reflect light to the human eye, and with the manner in which it does this transmitting or reflecting. For example, a transparent material that can be seen through without distorting the transmitted image is said to have good optical properties.

ORGANIC CHEMISTRY: Branch of chemical science that deals with the hundreds of thousands of known chemical molecules containing a large percentage of the element carbon in their make-up.

PHENOLICS: A generalized term that refers to all thermosetting resins of the phenol formaldehyde type, even though they may be made from different members of the phenolic family and the formaldehyde family.

PLASTICS: As used in this book refers to a synthetic polymerized resin; man made and not found in nature.

PLASTIC MEMORY: Tendency of a number of the thermoplastic resins to return to their original form, after being reheated, when they have previously been formed or distorted when hot.

PLASTICIZER: Any substance which is added to a plastics formulation to preserve a given degree of softness or plasticity in the material, even after it has set, due to cooling or chemical action.

POLYESTER RESINS: A large group of syrupy materials that can be made to polymerize into hard resins by adding a catalyst. Most fiber glass laminates are made using polyester resins.

POLYETHYLENE: Highly useful and very popular plastic made by causing ethylene

Glossary of Terms

gas to polymerize under suitable conditions.

POLYMER: Polymer refers to a molecule made up of many (poly) parts (mer). See monomer.

POLYMERIZATION: Chemical process that causes numerous molecules to join together to form a hugh molecule or polymer.

POT LIFE: Length of time that a resin will remain liquid or plastic before setting, after the catalyst and accelerator have been added.

POTTING: Process of encasing some object (such as a transformer coil) in a block of plastic by causing the plastic to set after the object has been suspended in it.

PROPERTIES: Properties of a substance may be physical, such as hardness, transparency, solidity, etc; or they may be chemical, such as being acid, or able to react with water, etc. The sum of the properties of any substance enable us to identify that substance and distinguish it from other substances.

RADICAL: Group of atoms acting as chemical unit during many types of chemical reactions.

REFRACTION: Measure of the angle through which a ray of light is bent when passing from one transparent substance into another of different light-transmitting properties.

RESIN: Highly polymerized substance, either of natural or synthetic origin that resembles ordinary resin in its characteristics. Resin and shellac are natural resins; plastics are synthetic resins.

SET: Set or setting, when used in speaking of plastics, means the act of hardening from the plastic condition. It usually signifies the chemical reaction that takes place during final heat treatment of thermosetting plastics.

STATIC ELECTRICITY: An electric charge that forms on the surface of typical nonconductors when they are rubbed or subjected to the proper kind of friction treatment.

STRUCTURAL STRENGTH: Measure of the resistance of any structure or form to crushing or distorting forces. For example a steel ball has greater structural strength than a table tennis ball.

SYNTHETIC: Any material that is built up by processes in the laboratory, set up and controlled by man rather than natural forces. For example, methanol (wood alcohol) made by combining carbon monoxide and water is synthetic. The same chemical obtained by distilling creosote, obtained by heating wood in the absence of air, is a natural form of methanol.

TAPPING: Cutting threads on the inside of a cylindrical hole.

TENSILE STRENGTH: A measure of the ability of a material to resist being pulled apart. A steel cable has high tensile strength.

THERMO: Of or pertaining to heat. A Thermometer is a heat measuring instrument - a thermoplastic becomes plastic under the influence of heat.

THERMOPLASTIC: Material that becomes soft or plastic when heated. Paraffin wax is a thermoplastic material.

THERMOSETTING: A substance that becomes a solid, or more hard, under the influence of heat.

THIXATROPIC: A substance that has a semi-solid, or jelly like form until it is agitated, then becomes a liquid. The liquid again sets to a jelly on standing.

THREADING: Process of cutting threads on the outside of a cylindrical rod, as in making bolts and studs from steel rods.

TRANSLUCENT: Ability of a substance to allow light to pass through it without transmitting a recognizable image.

TRANSPARENT: Ability not only to let light through, but transmit a visible, recognizable image.

UREA RESINS: Synthetic plastic materials made by the reaction between formaldehyde and urea. A nitrogen containing organic material that is readily manufactured from calcium cyanamid.

INDEX

A

Acrylic plastics, 13, 18
   internal carving of, 230 to 240 incl.
   lighting characteristics of, 21
   machining, 37
   painting, 240
   projects from, 55
   working with, section, 31
Antistatic treatment for plastics, 52

B

Band saw, using to cut plastic, 35
Bank, coin, 92, 93
Belting, heavy, used to make flexible form, 47
Bins, storage for Acrylics, 32
Bird feeder, 162, 163
Blending granules, 238
Board, cribbage, 185, 186, 187
Boat, covering with Fiberglas, 242
Bonding plastics, 45
Bonding, solvent, 48
Bowl, Fiberglas-resin serving, 248
Box, candy, blue and white, 188, 189
   carved-lid jewel, 142, 143
   cigarette, 224, 225, 226
   jewel, 190, 191
   jewel, cradled, 195, 196, 197
   jewel, sliding lid, 200, 201, 202
   piano-shaped jewel, 227, 228, 229
   powder, imported, 237
   powder with octagon base, 88, 89
   swept-corner, 176, 177
Bracelets, 68, 69
Bud vase, 106, 107
Bud vase, clef, 136, 137, 138
Buffing plastics, 41

C

Cake server, 58, 59
Candle holders, 100, 101, 114, 115
   fluted, 108, 109
Candy box, blue and white, 188, 189
Candy dish, basket type, 102, 103
   swan-shaped, 170, 171, 172
Candy, nut tray, wheelbarrow, 130, 131, 132
Carving roses, 234
Carving thin stock, 238
Casein, 30
Case, overnight, 250, 251, 252
Castolite casting, 260
Cast Phenolics, 260
Catalyst, 12
Cattail centerpiece, 182, 183, 184
Cellulosics, 14, 29
Cementing plastics, 45
Centerpiece, cattail, 182, 183, 184
Chair, modern bucket type, 253, 254
Checkerboard, 198, 199
Cigarette box and dispenser, 224, 225, 226
Cigarette lighters, 139, 140, 141
Circular saw, using to cut plastic, 34, 35
Cleaning plastics, 52
Clock, stand for electric, 154, 155
Coin bank, fish-shaped, 92, 93
Coin holders, 104, 105
Compressed cube carving, 146
Compression molding, 255
Cribbage boards, 185, 186, 187
Cube carving, compressed, 146, 147
Cup, plastic with cover, 98, 99
Cutouts, Styrofoam, 72, 73
Cutting tool, controlling, 232

D

Designs, transferring to plastics, 33
Desk lights, 139, 140, 141
Desk set with letter clip, 86, 87
Dish, candy, 102, 103
   candy, swan, 170, 171, 172
   pad, hot, 160, 161
Display, fabricating Plexiglas, 20
Drill bit, sharpening for use with plastics, 39
Drilling in plastics, 38
Drills, internal carving, 230

special recommended, 233
Dyes, Acrylic, 237
  laminating, 239

E

Earring and pin set, 110, 111
Earring rack, harp-shaped, 122, 123
Electric clock, stand for, 154, 155
Epoxy resins, 18
Ethylene dichloride, 47
Extrusion molding, 257
Eye dropper, using when cementing
  plastic, 48

F

Feeder, bird, 162, 163
Fiberglas, 242
Fiberglas-Polyester resin chair, 253, 254
Fiberglas-resin serving bowl, 248
"Fido" letter and note holder, 80, 81
Filaments, plastic, 263
Fireplace scene, Styrofoam, 74, 75
Foredom flexible shaft machine, 230
Forms, split block and tube, 46
Free-form lamp, 213, 214, 215
Friction welding plastics, 51

G

Gavel, laminated, 222, 223
Glossary of terms, 264
Granules, blending, 238

H

Handee motor tool, 230
Harp-shaped earring rack, 122, 123
Heater, strip, 44
Heat forming plastics, 42
Heat welding plastics, 51
Hot dish pad, 160, 161
Hypodermic needle, using when
  cementing plastic, 48

I

Injection molding, 256
Inorganic substances, 10
Internal carving of Acrylics, 230 to 240 incl.
Internal carving, pin and
  earring set, 110, 111

J

Jewel box, black and white, 190, 191
  carved lid, 142, 143
  cradled, 195, 196, 197
  sliding lid, 200, 201, 202
Jewelry findings, 239
Jig saw, using to cut plastic, 36
Jigs, using, 51

K

Kit, internal carving, 231
Knick-kack shelf, 158, 159

L

Laminated gavel, 222, 223
  letter openers, 82, 83
Laminates, built-up, 258
  color, 259
Lamp, modern design TV, 210, 211, 212
  modern free-form, 213, 214, 215
  night or TV, 207, 208, 209
  pin-up, 133, 134, 135
  planter, 173, 174
  plastic rod, 216, 217
  table, 192, 193, 194
  three-cone table, 204, 205, 206
  torchere, 218, 219
Lampshades from Polyplastex, 76, 77
Lapel pins, 203
Lathe tool ground for turning
  plastic, 36
Lazy Susan, 220, 221
Letter and note holder, 80, 81
Letter openers, laminated, 82, 83
  plaited, 96, 97
Lighters, cigarette, 139, 140, 141
  cube, 112, 113
Lighting characteristics of Acrylics, 21
Lignin, 30

M

Magazine rack, 116, 117
Melamines, 16
Metallizing plastics, 263
Metals, applying plastics to, 247
Methylene dichloride, 47
Methyl methacrylate, 48
Methyl methacrylate resin,
  formula for, 19
Mirror, vanity, 156, 157
Moslo molding machine, 256
Musical clef shelf, 126, 127

Index

N

Name plates, raised letter, 152, 153
Napkin holder, 78, 79
   weighted, 124, 125
Natural resins, 30
Natural resins, casein, 30
   lignin, 30
   shellac, 30
Night or TV lamp, 207, 208, 209
Nut tray, 60, 61
Nylon, 25, 26

O

Organic substances, 10
Ornaments, Styrofoam, 64, 65
Overnight case, 250, 251, 252

P

Painting Acrylics, 240
Pen sets, 150, 151
Pencil holder, 70, 71
Perfume
   atomizer stand, heart-shaped, 168, 169
   bottle holders, 148, 149
   vial holders, 90, 91
Phenolics, 14
   cast, 236
Piano shaped cigarette dispenser
   or jewel box, 227, 228, 229
Picture frame, laced, 118, 119
Pie server, 58, 59
Pin and earring set, 110, 111
Pins, lapel, 203
Pipe holder, 128, 129
Planter lamp, 173, 174
Plastic, Acrylic, working with, 31
   casting, 259
   filaments, 263
   foams, 261
   internal carving, 230 to 240 incl.
   lamination, wet lay-up method, 244
   moldings, types of, 255
   Polyamide, 25
   Polyethylene, 24
   rod lamp, 216, 217
   sawing, 34, 35, 36
   storing Acrylic, 31
   synthetic, first, 9

Plastics, 13, 18
   antistatic treatment for, 52
   applying to metals, 247
   buffing, 41
   cementing and bonding, 45
   classifying, 13
   cleaning, 52
   commercial methods of fabricating
     section, 255 to 263 incl.
   drilling in, 38
   Epoxy, 18
   facts about, 9
   friction welding, 51
   heat forming, 42
   heat welding, 51
   machining, 37
   Melamines, 16
   metallizing, 263
   Nylon, 25, 26
   Phenolics, 14
   polishing, 41
   Polyesters, 17
   Polyfluoro Hydrocarbons, 28
   Polystrene, 23
   Polyvinylidene Chloride, 27
   projects from, 55
   sanding, 39
   sculpturing, 241
   Silicone, 18
   solvent polishing, 42
   Thermoplastic, 13
   Thermosetting, 13
   threading and tapping, 39
   types in common use, 14
   Ureas, 15
   Vinyl, 27
   waxing, 52
Plastic-like materials, Cellulosics, 29
Plastic-like materials, section, 29
Polishing plastics, 41
Polyamide plastics, 25
Polyesters, 17
Polyester resins and reinforced
   laminates section, 242 to 254 incl.
Polyethylene plastic, 24
Polyfluoro Hydrocarbons, 28
Polymerization, 11
Polyplastex, lamp shades from, 76, 77
Polystrene plastic, 23
Polyvinylidene Chloride plastics, 27

Powder box, octagon base, 88, 89
Project section, 55

R

Rack, magazine, 116, 117
Rack, tie, 94, 95, 166, 167
    towel, 144, 145
Raised letter name plates, 152, 153
Resin, methacrylate, 19
Resins, natural, 14
    Polyester, 242, 244
Rings, finger, 66, 67
Roses, carving, 234
Rubber stamps, stand for, 164, 165
Rubbers, synthetic, 28

S

Salt and pepper shakers, 180, 181
Sanding plastics, 39
Sawing plastic, 34, 35, 36
Sculpturing plastics, 241
Shakers, salt and pepper, 180, 181
Sharpening drill bit for plastics, 39
Shelf, knick-knack, 158, 159
    musical clef, 126, 127
Shellac, 30
Silicones, 18
Solvent bonding, 48
Solvent polishing plastics, 42
Stamps, rubber, stand for, 164, 165
Steel burrs, 230
Stokes molding machine, 257
Strip heater, using, 44
Styrofoam, 261
    cutouts, 72, 73
    fireplace scene, 74, 75
    ornaments, 64, 65
Swan, white, candy dish, 170, 171, 172
Swept-corner box, 176, 177
Synthetic plastic, first, 9
Synthetic rubbers, 28

T

Tenite tool handles, 120, 121
Terms, glossary of, 264
Thermoplastic plastics, 13
    resins, 14, 18
Thermosetting plastics, 13
    resins, 14
Thin stock, carving, 238
Threading and tapping plastics, 39
Three-cone table lamp, 204, 205, 206
Three tier tray, 178, 179
Tie rack, 94, 95
    of clear plastic, 166, 167
Tie slides, 62, 63
Tool handles, Tenite, 120, 121
Tooth brush holder, 56, 57
Torchere lamp, 218, 219
Towel rack, wall, 144, 145
Tray, forming with split block and
    tube forms, 46
    lazy susan, 220, 221
    nut, 60, 61
    three tier, 178, 179
TV lamp of modern design, 210, 211, 212

U

Ureas, 15

V

Vanity mirror, 156, 157
Vase, clef bud, 136, 137, 138
    bud, 84, 85, 106, 107
Vinyl plastics, 27
    Trichloride, 47

W

Waxing plastics, 52
Wheelbarrow, candy, nut tray, 130, 131